T0214745

# Texts and Readings in Mathematics

## Volume 55

The **Texts and Readings in Mathematics** series publishes high-quality textbooks, research-level monographs, lecture notes and contributed volumes. Undergraduate and graduate students of mathematics, research scholars and teachers would find this book series useful. The volumes are carefully written as teaching aids and highlight characteristic features of the theory. Books in this series are co-published with Hindustan Book Agency, New Delhi, India.

More information about this series at https://link.springer.com/bookseries/15141

Sebastian M. Cioabă · M. Ram Murty

# A First Course in Graph Theory and Combinatorics

## Second Edition

 HINDUSTAN
BOOK AGENCY

 Springer

Sebastian M. Cioabă
Department of Mathematical Sciences
University of Delaware
Newark, DE, USA

M. Ram Murty
Department of Mathematics
Queen's University
Kingston, ON, Canada

ISSN 2366-8717          ISSN 2366-8725  (electronic)
Texts and Readings in Mathematics
ISBN 978-981-19-1362-4          ISBN 978-981-19-0957-3  (eBook)
https://doi.org/10.1007/978-981-19-0957-3

This work is a co-publication with Hindustan Book Agency, NewDelhi, licensed for sale in all countries in electronic form, in print form only outside of India. Sold and distributed in print within India by Hindustan Book Agency, P-19 Green Park Extension, New Delhi 110016, India.
ISBN: 978-81-95196-18-0

This Springer imprint is published by the registered company Springer Nature Singapore Pte Ltd.
The registered company address is: 152 Beach Road, #21-01/04 Gateway East, Singapore 189721, Singapore

*This book is dedicated to the memory of Professor Gian-Carlo Rota, mathematician, philosopher, and teacher.*

# Preface to Second Edition

*The visible has its roots in the unseen*
*And each invisible hides what it can mean*
*In a yet deeper invisible, unshown.*
Sri Aurobindo, Discoveries of Science III

This text arose from courses given by the authors at Queen's University and the University of Delaware, over a period of about 20 years. The urgent need to introduce the topic of graph theory and combinatorics to senior undergraduate students and beginning graduate students in a single semester course is quite challenging. The time constraint alone forces one to make a sound selection of topics that would help the student in their future careers, whatever these may be, ranging from a research-oriented academic environment to a more mercantile application of this knowledge in the world of business and industry. So, one must strike a balance between theory and applications. At the same time, one has to give a basic understanding that covers the essential ideas so that the student may probe further, if needed. Indeed, in our current digital age, people find new applications of graph theory almost every day and the student is in a good position to apply this knowledge to these new challenges.

Since the first edition, there have been many new advances in the field. It is impossible to include a discussion of all of these achievements. We have singled out one in the final chapter. This concerns the existence of Ramanujan graphs of every degree. We indicate briefly these new developments and invite the student to study further. We have also taken this opportunity to correct typos and errors in the first edition. We feel this new edition will be ideal for any student seeking a quick introduction to the subject.

The authors thank their postdocs and students and are grateful to Kathryn Beck, Kris Hollingsworth, Seoyoung Kim, Mark Leadingham, Nathaniel Merrill, Mutasim

Mim, Danny Rorabaugh, Jake Sitison, Paul Steller, Nathan Thom, and Alex Wilton for reading various versions of the manuscript and their feedback.

Newark, USA                                                                Sebastian M. Cioabă
Kingston, Canada                                                              M. Ram Murty
August 2021

# Preface to First Edition

*The butterfly counts not months but moments and has time enough.*

Rabindranath Tagore

The concept of a graph is fundamental in mathematics, since it conveniently encodes diverse relations and facilitates combinatorial analysis of many complicated counting problems. In this book, we have traced the origins of graph theory from its humble beginnings of recreational mathematics to its modern setting for modelling communications networks, as is evidenced by the World Wide Web used by many internet search engines.

This book is an introduction to graph theory and combinatorial analysis. It is based on courses given by the second author at Queen's University at Kingston, Canada, between 2002 and 2008. These courses were aimed at students in their final year of their undergraduate program. As such, we believe this text is very suitable for a first course on this topic.

Graph theory and combinatorics interact well with other branches of mathematics like number theory, algebraic topology, algebraic geometry, and representation theory, as well as other sciences. For instance, Ramanujan graphs and expander graphs have gained prominence with applications to the construction of optimal communication networks. Thus, we have included a chapter on this important emerging theme at the end of the book.

There are many books on graph theory and combinatorics. What makes this book unique is that we have tried to make it suitable for self-study. Students and non-experts should be able to work through the book at their own pace without an instructor. Hints to the exercises have also been included to facilitate this study.

The book can be also used for a course at the college level. The material can be easily covered in two semesters. Instructors may find it easy to highlight the graph-theoretic aspects in one course, and the combinatorial aspects in another. For instance, Chapters 1, 3, 4, 5, 6,8, 10, 11, 12 can be used for a one semester course in graph theory. Chapters 2, 3, 6, 7, and 9 can comprise a short semester course in combinatorics.

At the end of this book, we present a brief list of books and papers that give more details about some topics discussed here.

April 2009                                                                 Sebastian M. Cioabă
                                                                            M. Ram Murty

# Contents

# About the Authors

**Sebastian M. Cioabă** is Professor at the Department of Mathematical Sciences, University of Delaware, Newark, USA. After his undergraduate studies in mathematics and computer science at the University of Bucharest, Romania, he obtained his Ph.D. in Mathematics at Queen's University at Kingston, Canada. Following postdocs at UC San Diego and the University of Toronto, Sebastian started his position at the University of Delaware in 2009. His research interests are in spectral graph theory, algebraic combinatorics, and their connections and applications to other areas of mathematics and science. He is on the editorial board of several journals including *Discrete Mathematics, Linear Algebra and its Applications*, and *Electronic Journal of Linear Algebra*. He has organized several conferences in algebraic combinatorics and spectral graph theory. Sebastian has supervised or co-supervised 6 Ph.D. students, 2 M.Sc. students, 3 undergraduate senior theses, and over 20 summer research students. He has published more than 65 papers, and his research has been supported by NSF, NSA, NSERC, Simons Foundation, IDex Bordeaux, and Japan Society for Promotion of Science.

**M. Ram Murty** is Queen's Research Chair and A.V. Douglas Distinguished University Professor at Queen's University, in Kingston, Ontario, Canada. He obtained his Ph.D. from Massachusetts Institute of Technology, USA, in 1980 and subsequently held positions at the Institute for Advanced Study in Princeton, Tata Institute for Fundamental Research in Mumbai, and McGill University in Montreal. He has authored more than 250 research papers and written more than a dozen mathematical textbooks. His monograph, *Non-vanishing of L-functions and Applications*, written jointly with Prof. V. Kumar Murty, won the 1996 Balaguer Prize. Ram is Fellow of the Royal Society of Canada, Fellow of the American Mathematical Society, and Fellow of the Indian National Science Academy, India. He also teaches Indian philosophy at Queen's University and has authored *Indian Philosophy: An Introduction*, published by Broadview Press.

# Chapter 1
# Basic Graph Theory

## 1.1 Königsberg Bridges Problem

Cioabă Graph theory may be said to have begun with the 1736 paper by the Swiss mathematician Leonhard Euler (1707–1783) devoted to the Königsberg bridges problem. In the town of Königsberg (now Kaliningrad in western Russia), there were four land masses and seven bridges connecting them as shown in Fig. 1.1. The challenge was to leave home and to traverse each bridge exactly once and return home.

While Euler did not construct a *graph* per se, his argument is equivalent to considering a graph representation of the problem (see Fig. 1.2).

The two sides of the river and the two islands are represented by *vertices* or *points* in the plane. They are joined by an *edge* if there is a bridge between them.

The Königsberg bridges problem reduces to determining the existence of a *circuit* through the graph which traverses each edge only once. Euler reasoned that if one were able to go through every bridge exactly once and return to the starting point, the *degree* of each vertex (the number of edges or bridges coming out of any vertex or land mass) must be even (see Fig. 1.3).

In the Königsberg bridge graph, the degree of each vertex is odd and hence, no such circuit exists. This example illustrates many of the basic notions of graph theory, which we take up in the next section.

## 1.2 What Is a Graph?

A **graph** $X$ is a pair $(V, E)$ of a set of **vertices** $V$ and **edges** $E$ that associates with each edge two vertices (not necessarily distinct) called its **endpoints**. A **loop** is an edge whose endpoints are the same. **Multiple edges** are edges having the same pair of endpoints. A graph is called **simple** if it has no loops or multiple edges.

© Hindustan Book Agency 2022
S. M. Cioabă and M. R. Murty, *A First Course in Graph Theory and Combinatorics*,
Texts and Readings in Mathematics 55, https://doi.org/10.1007/978-981-19-0957-3_1

**Fig. 1.1** The bridges of
Königsberg

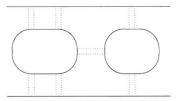

**Fig. 1.2** A graph
representation of the
Königsberg bridges

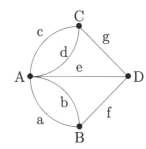

**Fig. 1.3** A vertex in an
Eulerian cycle

**Fig. 1.4** A graph with 5
vertices and 8 edges

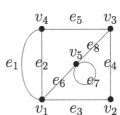

The graph $X$ in Fig. 1.4 has a loop $e_7$ at vertex $v_5$ and $e_1$ and $e_2$ are multiple edges.
It is not a simple graph and has

$$V = \{v_1, v_2, v_3, v_4, v_5\},$$
$$E = \{e_1, e_2, e_3, e_4, e_5, e_6, e_7, e_8\}.$$

A graph is said to be **finite** if $V$ and $E$ are finite. Unless otherwise specified, we
will be treating only finite graphs in this book. In some instances, we use the notation

**Fig. 1.5** The complete
graph $K_5$

$V(X)$ for $V$ and $E(X)$ for $E$ to avoid confusion between various vertex sets and edge
sets.

The graph $Y$ in Fig. 1.2 has $V(Y) = \{A, B, C, D\}$ and $E(Y) = \{a, b, c, d, e, f, g\}$,
does not contain any loops, but it contains multiple edges such as $a$ and $b$ or $c$ and
$d$. Therefore, it is not a simple graph.

When $x$ and $y$ are endpoints of an edge $e$, we say they are **adjacent** or are
**neighbours**. We also say that the vertex $x$ (or $y$) is **incident** with the edge $e$. The
**degree** (also known as the **valency** or the **valence**) of a vertex is the number of
edges that contain it. We denote the degree of the vertex $x$ by $d(x)$. Note that a loop
containing $x$ contributes 2 to $d(x)$ while any other edge containing $x$ contributes 1 to
its degree. A vertex $x$ is said to be **odd** or **even** according to $d(x)$ being odd or even.
The graph $Y$ in Fig. 1.2 has $d(A) = 5$ and $d(B) = d(C) = d(D) = 3$. The reader
can find the degree sequence of the graph $X$ in Fig. 1.4.

Our first theorem of graph theory connects the sum of the degrees to the number
of edges and is usually called the Handshaking lemma, as it can be interpreted as
stating that the sum of the number of handshakes people do at some meeting equals
twice the total number of handshakes.

**Theorem 1.2.1** *For a finite graph* $X = (V, E)$,

$$\sum_{x \in V} d(x) = 2|E|.$$

***Proof*** We use double counting, which means that we will count a certain quantity
in two different ways. Consider the collection of all ordered pairs $(x, e)$, where $x$ is
a vertex, $e$ is an edge with $x$, and $e$ being incident. If we count the number of such
ordered pairs by starting with the first coordinate, we get the answer

$$\sum_{x \in V} |\{e : e \in E, x \text{ incident to } e\}| = \sum_{x \in V} d(x).$$

If we now count the same collection of ordered pairs by starting with the second
coordinate, we get the answer $2|E|$. This proves the result.                        ∎

The next result follows by considering the above equation mod 2.

**Corollary 1.2.2** *In any finite graph, the number of odd vertices is even.*

For any integer $n \geq 1$, the **complete graph** $K_n$ on $n$ vertices is the simple graph
on $n$ vertices where any two distinct vertices are adjacent (Fig. 1.5).

**Fig. 1.6** The complete
bipartite graph $K_{3,4}$

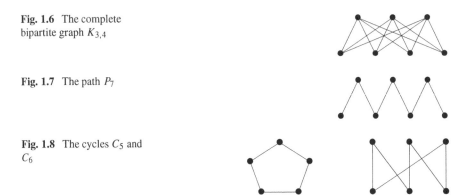

**Fig. 1.7** The path $P_7$

**Fig. 1.8** The cycles $C_5$ and
$C_6$

Given a graph $X = (V, E)$, **the complement** $X^c$ of the graph $X$ is the graph with
the same vertex set as $X$, where $x$ and $y$ are adjacent in $X^c$ if and only if $x$ and $y$ are
not adjacent in $X$. The complement $K_n^c$ of $K_n$ is usually called **the empty graph on**
$n$ **vertices**.

A **clique** in $X$ is a subset of vertices that are pairwise adjacent. **The clique number**
$\omega(X)$ of $X$ is defined as the largest order of a clique in $X$. A graph with $n$ vertices
has clique number $n$ if and only if it is $K_n$. Also, a graph with $n$ vertices has clique
number 1 if and only if it has no edges and is the empty graph $K_n^c$.

An **independent set** or **stable set** is a subset of vertices, no two of which are
adjacent. **The independence number** $\alpha(X)$ is the largest order of an independent
set of $X$. It is easy to see that $(X^c)^c = X$ and that a clique in $X$ is the same as an
independent set in its complement $X^c$. Therefore, $\omega(X) = \alpha(X^c)$.

A graph is called **bipartite** if the vertex set is the union of two (possibly empty)
disjoint independent sets. The **complete bipartite graph** $K_{r,s}$ is the bipartite graph
whose vertex set is a disjoint union of two independent sets of size $r$ and $s$ with every
vertex in the first set being adjacent to every vertex in the second set (Fig. 1.6).

For $n \geq 1$, the **path on** $n$ **vertices** $P_n$ is the simple graph on $n$ vertices $x_1, \ldots, x_n$
whose edges are $\{x_j, x_{j+1}\}$, for $1 \leq j \leq n$ (Fig. 1.7).

When $n \geq 3$, the **cycle on** $n$ **vertices** or the $n$-**cycle** $C_n$ is the simple graph on
$n$ vertices $y_1, \ldots, y_n$ whose edges are $\{y_j, y_{j+1}\}$, for $1 \leq j \leq n$ where we interpret
$y_{n+1}$ as $y_1$. Note that every path is a bipartite graph, while a cycle is a bipartite graph
if and only its number of vertices is even (Fig. 1.8).

Until 1976, one of the most famous unsolved problems of mathematics was the
Four Colour Conjecture. This conjecture says that every map can be properly coloured
using only four colours, where a proper colouring means that no two adjacent regions
should be coloured the same. This problem has a long history. It appears in a letter
of October 23, 1852, from the English mathematician Augustus de Morgan (1806–
1871) to the Irish mathematician William Rowan Hamilton (1805–1865) and was
asked by one of de Morgan's students Frederick Guthrie who later attributed to his
brother Francis Guthrie. In 1878, the English mathematician Arthur Cayley (1821–
1895) announced the problem to the London Mathematical Society. In 1879, the

English mathematician Alfred Kempe (1849–1922), published a *proof*. In 1890, the English mathematician Percy John Heawood (1861–1955) indicated there was a gap in Kempe's proof and gave a simple proof that *five colours suffice*, that is the Five Colour Theorem. We will prove this result later in Chap. 10. The Four Colour Conjecture was finally solved in 1976 by the American mathematician Kenneth Appel (1932–2013) and the German mathematician Wolfgang Haken using extensive computer verification. For some, this is not satisfying and the search is still on for a more conceptual and clearer solution.

Graphs arise in diverse contexts, and many *real world* problems can be formulated graph-theoretically. An important problem that arises in practice is the following scheduling one. Suppose we have $n$ timetable slots in which to schedule $r$ classes. We want a timetabling so that no student has a conflict. We can create a graph with $r$ vertices, each vertex denoting a class. We join two vertices if they have a common student, meaning that they cannot be scheduled in the same time slot. We want to *colour* the graph using $n$ colours so that no two adjacent vertices have the same colour. The **chromatic number** $\chi(X)$ of a graph $X$ is the minimum number of colours needed to color the vertices so that no two adjacent vertices have the same colour. In the scheduling problem, $\chi(X)$ will equal the smallest number of time slots for the schedule.

Bipartite graphs arise in job assignment questions. Suppose we have $m$ jobs and $n$ people, but not all people are qualified to do the job. Can we make job assignments so that all the jobs are done? Each job is filled by one person, and each person can hold at most one job. Thus, we can create a bipartite graph consisting of $n$ people and $m$ jobs and join a person to a job if the person can do the respective job.

Sometimes, we can assign *weights* to edges and this facilitates discussion of *routing problems*. Suppose we have a road network. The edges will correspond to road segments, and the weights can be the distance between the two points. Questions concerning the shortest path from point $a$ to point $b$ can be formulated in terms of finding the geodesic between $a$ and $b$.

## 1.3 Mathematical Induction

In many proofs in graph theory and combinatorics, we require the principle of **mathematical induction**, which we briefly recall below. Suppose we have a sequence of propositions $(P(n))_{n \geq 1}$ indexed by the natural numbers that we would like to prove. We begin by verifying that $P(1)$ is true. If we can prove that for any natural number $k$, $P(k)$ implies $P(k + 1)$ (or $P(k) \Rightarrow P(k + 1)$ for short), then $P(n)$ is true for any $n \geq 1$. A common analogy for a mathematical induction proof is imagining a sequence of dominoes representing our propositions $P(1), P(2), \ldots, P(k), P(k + 1), \ldots$. Proving that $P(1)$ is true ensures that the first domino will fall. Proving that $P(k) \Rightarrow P(k + 1)$ for any $k \geq 1$ ensures that all the dominoes fall as follows: $P(2)$ goes down first since $P(1)$ is true and $P(1) \Rightarrow P(2)$, $P(3)$ falls next since $P(2)$ is true and $P(2) \Rightarrow P(3)$ and so on.

We give below a simple illustration of this principle. Notice that

$$1^3 + 2^3 = 9 = 3^2$$
$$1^3 + 2^3 + 3^3 = 6^2$$
$$1^3 + 2^3 + 3^3 + 4^3 = 10^2,$$

a pattern first noticed by Aryabhata in the fifth-century India. He showed that for any natural number $n$, the following identity is true:

$$1^3 + 2^3 + \cdots + n^3 = \left(\frac{n(n+1)}{2}\right)^2.$$

He did this essentially by the principle of mathematical induction. To see the details, denote by $s_n$ the left-hand side of the equation above. Our proposition $P(n)$ is the statement that $s_n = \left(\frac{n(n+1)}{2}\right)^2$. For $n = 1$, $P(1)$ is true since $s_1 = 1^3 = 1$ and $\left(\frac{1(1+1)}{2}\right)^2 = 1$. Let $k$ be a natural number. Assume that $P(k)$ is true, namely that $s_k = \left(\frac{k(k+1)}{2}\right)^2$. We want to prove that $P(k+1)$ is true, meaning that $s_{k+1} = \left(\frac{(k+1)(k+2)}{2}\right)^2$. To do this, first note that $s_{k+1} = s_k + (k+1)^3$. By the induction hypothesis, $s_k = \left(\frac{k(k+1)}{2}\right)^2$, so we get that

$$s_{k+1} = \left(\frac{k(k+1)}{2}\right)^2 + (k+1)^3 = \left(\frac{(k+1)(k+2)}{2}\right)^2,$$

as required.

In some situations, it is more convenient to use the method of strong induction. The scenario is the same as before: we try to prove that a sequence of mathematical statements $(P(n))_{n \geq 1}$ are true. If we can prove that $P(1)$ is true (the base case is the same as in the case of usual induction) and also that for any $k \geq 1$, $P(\ell)$ true for any $1 \leq \ell \leq k$ implies that $P(k+1)$ is true, then this will imply that $P(n)$ is true for any $n \geq 1$. The name *strong* refers to the hypothesis of the induction step, namely, $P(\ell)$ true for any $1 \leq \ell \leq k$, which is stronger than the hypothesis of the usual induction method which is that $P(k)$ is true.

We will derive now two important graph theory theorems by the method of mathematical induction. The first result concerns the parity of cycles in graphs. The second theorem characterizes Eulerian graphs and will be given in the next section.

A **walk** in a graph $X = (V, E)$ is an alternating sequence

$$x_0, e_1, x_1, \ldots, e_k, x_k,$$

of vertices $x_0, \ldots, x_k$ and edges $e_1, \ldots, e_k$ such that for $1 \le j \le k$, the edge $e_i$ has endpoints $x_{i-1}$ and $x_i$. We sometimes refer to such walk as a $(x_0, x_k)$-walk to indicate the first vertex $x_0$ and the last vertex $x_k$ of the walk. The **length** of this walk is defined as $k$, the number of edges in it. A walk is **odd** or **even** according as the length of the walk is odd or even, respectively. A $(x_0, x_k)$-walk is called **closed** if $x_0 = x_k$.

For the graph in Fig. 1.2, the sequence

$$A, a, B, b, A, c, C, g, D, g, C, c, A$$

is a closed walk of length 6, while the sequence $B, f, D, g, C, e, B$ is not a walk (why?).

A **trail** is a walk with no repeated edges. Our previous example of a walk is not a trail. A **path** is a walk with no repeated vertices. Thus, any path is a trail, but not any trail is a path. For the graph in Fig. 1.2, the sequence $C, d, A, c, C, g, D$ is a trail that is not a path. The sequence $B, f, D, e, A, c, C$ is a path.

A **circuit** is a closed trail. A **cycle** is a closed trail that does not repeat any vertex except for the first vertex and the last vertex. A cycle of length one is the same as a loop, while a cycle of length two consists of two edges having the same endpoints. We speak of odd or even paths, trails, cycles, and circuits, according to their lengths are odd or even, respectively.

**Lemma 1.3.1** *Every closed odd walk contains an odd cycle.*

***Proof*** We use strong induction on the length $\ell$ of the closed walk $W$. For $\ell = 1$, a closed walk of length one is also a cycle of length one, so there is nothing to prove. Now suppose that the assertion has been established for odd walks of length $\ell - 1$ or less. If $W$ has no repeated vertices except the first and the last one, then $W$ itself is a cycle, and we are done. Otherwise, we may suppose that some vertex $x$ is repeated in $W$. We can think of the walk as starting from $x$ and view $W$ as two $(x, x)$-walks $W_1$ and $W_2$ (say). The length of $W$ is the sum of the lengths of $W_1$ and $W_2$. As the length of $W$ is odd, one of $W_1$ or $W_2$ must have an odd length which is necessarily smaller than the length of $W$. By the induction hypothesis, this odd walk must have an odd cycle. Hence, $W$ will contain an odd cycle. ∎

Note that the previous statement is not true if odd is replaced by even, as a closed even walk using exactly one edge does not contain any (even) cycles.

A graph $X = (V, E)$ is said to be **connected** if, for any $x \ne y \in V$, there exists a $(x, y)$-walk. The following result implies this is the same as for any $x \ne y \in V$, there exists a $(x, y)$-path.

**Lemma 1.3.2** *Let $X = (V, E)$ be a graph and $x \ne y \in V$. There exists a $(x, y)$-walk if and only if there exists a $(x, y)$-path.*

***Proof*** A path is a walk. For the other direction, one can use strong induction on the length of the walk and an argument similar to the previous lemma. ∎

**Fig. 1.9** An illustration of
the proof of Lemma 1.4.2

$u$                                     $v$                                            $P$

Given a graph $X = (V, E)$, **a subgraph** $Y = (W, F)$ is a graph obtained by
deleting some vertices and some edges (possibly none) from $X$. The subgraph $Y$ is
called **induced** or **induced by** $W$ if it is obtained from $X$ by deleting only vertices
(possibly none). With this definition, a graph $X$ is an induced subgraph of itself.

Define the following binary relation on the vertex set $V$ of a graph $X$:

$$x \sim y \text{ if } x = y \text{ or there exists an } (x, y) - \text{path.}$$

This is an equivalence relation, meaning that it is reflexive, symmetric, and transitive.
The subgraphs of $X$ induced by its equivalence classes are called **the components** or
**the connected components** of $X$. A graph is connected if and only if it has exactly
one component. Equivalently, it is connected if and only if, for any two vertices
$x \neq y$, there is a $(x, y)$-path.

## 1.4    Eulerian Graphs

A graph is called **Eulerian** if it has a closed trail containing all edges. Note that a
graph is Eulerian if and only any graph obtained by adding or removing any isolated
vertices to or from it is Eulerian.

The main result of this section is the following characterization of Eulerian graphs.

**Theorem 1.4.1** (Euler-Hierholzer) *A graph without isolated vertices is Eulerian if
and only if it is connected and all vertices have even degree.*

It seems that Euler did not give complete proof in his 1741 paper. The first complete
published proof was given by Karl Hierholzer (1840–1871) in a posthumous article
in 1873. The graph we drew in Fig. 1.2 to model the problem did not appear in print
until 1894. Before we prove the theorem above, we need the following lemma.

**Lemma 1.4.2** *If every vertex of a graph $X$ has degree 2 or more, then $X$ contains a
cycle.*

**Proof** Let $P$ be a maximal path in $X$. Let $u$ be an endpoint of $P$. Since $P$ is maximal,
every neighbour of $u$ must already be a vertex of $P$ otherwise, $P$ can be extended.

Since $u$ has degree at least 2, it has a neighbour $v$ in $P$ via an edge not in $P$ (see
Fig. 1.9). The edge $uv$ completes a cycle with the portion of $P$ from $v$ to $u$.    ∎

We now give the proof of Euler-Hierholzer Theorem.

**Fig. 1.10**  A connected
graph with even degrees

***Proof*** *(Theorem* 1.4.1*)* For the necessity, if $X$ is an Eulerian graph without isolated vertices, then by our previous arguments in Sect. 1.1, we deduce that the degree of each vertex of $X$ must be even. Also, for any vertices $x \neq y$, because $X$ has no isolated vertices and $d(x)$ and $d(y)$ are even, there must be at least two edges incident with $x$ and $y$, respectively. A closed trail containing all the edges of $X$ will therefore go through $x$ and $y$. Hence, there will be a $(x, y)$-walk. This proves that $X$ must be connected.

For sufficiency, we use strong induction on the number $m$ of edges of $X$. If $X$ has one edge, then $X$ consists of one vertex and one loop edge. Thus, $X$ is Eulerian. Let $m \geq 2$ now and consider a connected graph $X$, where each vertex has an even degree. Therefore, every vertex of $X$ has a degree of at least 2. By Lemma 1.4.2, $X$ contains a cycle $C$. Let $Y$ be the graph obtained from $X$ by deleting the edges of the cycle $C$. The graph $Y$ may not be connected as shown in Fig. 1.10 where deleting the edges of the cycle on four vertices leaves a graph on eight vertices whose components are two vertex disjoint cycles of lengths three and two isolated vertices.

Since $C$ has 0 or 2 edges at each vertex, each component of $Y$ is either a connected graph whose degrees are all even or is a trivial component consisting of a single vertex with no loops. By induction, each non-trivial component of $Y$ has an Eulerian circuit. We combine each such component with $C$ to get an Eulerian circuit of $X$ as follows. We pick a vertex on $C$ and start walking on the edges of $C$. Every time we encounter a non-trivial component of $Y$ while traversing $C$, we go through the Eulerian circuit of that component. Every time we meet an isolated vertex of $Y$, we continue our walk on $C$. In the end, we must return to the vertex where we started, and we have used each edge of $X$ exactly once. Hence, $X$ is Eulerian.  ∎

The notion of graph can be generalized to that of directed graph or **digraphs** (see Exercise 1.6.5) and one can prove analogues of the Handshaking Lemma (see Exercise 1.6.6) and the Euler-Hierholzer Theorem (see Exercise 1.6.7) in that realm.

## 1.5  Bipartite Graphs

Dénes König (1884–1944) studied in Budapest and Göttingen. His book *Theorie der endlichen und unendlichen Graphen—Theory of finite and infinite graphs* which appeared in 1936 is considered to be the first monograph in graph theory and contributed greatly to the growing interest in this subject.

Recall that a graph $X = (V, E)$ is called bipartite if $V$ is the union of two (possibly empty) disjoint independent sets. If $V = A \cup B$, $A \cap B = \emptyset$ and $A$ and $B$ are independent sets, we call the partition $V = A \cup B$ a partition of the vertex set of

$X$ into **the partite sets** of $X$. Note that for a bipartite $X$, the partition of $V$ into partite sets is not necessarily unique (consider the empty graph on four vertices, for example).

König obtained the following characterization of bipartite graphs.

**Theorem 1.5.1** (König) *A graph $X$ is bipartite if and only if it has no odd cycles.*

*Proof* It is not too hard to observe that a graph is bipartite if and only if each of its components is bipartite. Therefore, it suffices to prove the result above for connected graphs.

For necessity, if $X$ is a connected bipartite graph whose vertex set is partitioned into two independent sets $A$ and $B$, then observe that every walk alternates between these independent sets. For any closed walk, the return to the starting vertex must happen after traversing an even number of edges. Hence, $X$ has no odd closed walks, implying that it has no odd cycles.

For the sufficiency, let $X = (V, E)$ be a connected graph with no odd cycles. Let $x$ be a vertex of $X$. For each $y \in V$, denoted by $f(y)$ be the shortest length of a $(x, y)$-path. Since $X$ has no loops (why?), $f(x) = 0$. Also, $f(y) = 1$ for any neighbour $y$ of $x$. Because $X$ is connected, $f(y)$ is well-defined of every $y \in V$.

Let

$$A = \{y \in V : f(y) \text{ is even}\} \text{ and } B = \{z \in V : f(z) \text{ is odd}\}.$$

Clearly, $A$ and $B$ form a partition of $V$. Also, both $A$ and $B$ are independent sets. To see this, assume that $y \neq y' \in A$ are adjacent. From the definition of $A$, there is an even $(x, y)$-path and an even $(x, y')$-path. Combining these paths with the edge $yy'$, we obtain a closed odd walk. By Lemma 1.3.1, $X$ contains an odd cycle, contrary to assumption. By a similar argument, one can prove that $B$ is an independent set. Hence, $X$ is bipartite. ∎

We conclude this section with a simple result concerning the degrees of the vertices of a bipartite graph. This can be seen as the bipartite version of the Handshaking Lemma.

**Theorem 1.5.2** *If $X = (V, E)$ is a bipartite graph with partite sets $A$ and $B$, then*

$$\sum_{a \in A} d(a) = \sum_{b \in B} d(b) = |E|.$$

*Proof* Count all the ordered pairs in the set

$$\{(a, e) : x \in A, e \in E, a \text{ incident to } e\}.$$

If we start with the first coordinate, for each $a \in A$, there are exactly $d(a)$ edges $e$ that are incident with $a$. Summing up over all $a \in A$, we get the left-hand side in the equation above. If we start our count using the second coordinate, then each edge $e \in E$ is adjacent to one vertex $a \in A$ (and another $b \in B$). Therefore, our count will now be $|E|$. The proof that $\sum_{b \in B} d(b) = |E|$ is similar. ∎

## 1.6 Exercises

**Exercise 1.6.1** Is there a simple graph with 9 vertices and degrees

$$3, 3, 3, 3, 5, 6, 6, 6, 6?$$

**Exercise 1.6.2** Is there a bipartite graph of 8 vertices with degrees

$$3, 3, 3, 5, 6, 6, 6, 6?$$

**Exercise 1.6.3** Let $n$ be a natural number. Show that there exists a unique (up to isomorphism) graph on $n$ vertices whose degree sequence contains any natural number from 1 to $n - 1$.

**Exercise 1.6.4** Show that for any natural number $n \geq 2$,

$$1^2 + 2^2 + \cdots + n^2 = \frac{n(n+1)(2n+1)}{6},$$

$$1 + 3 + \cdots + (2n - 1) = n^2.$$

**Exercise 1.6.5** A **directed graph** (or **digraph**) is a graph $X = (V, E)$ together with a function assigning to each edge, an **ordered** pair of vertices. The first vertex is called the **tail** of the edge, and the second is called the **head**. To each vertex $x$, we let $d^+(x)$ be the number of edges for which $x$ is the tail and $d^-(x)$ the number for which it is the head. We call $d^+(x)$ the **outdegree** and $d^-(x)$ the **indegree** of $x$. Prove that

$$\sum_{x \in X} d^+(x) = \sum_{x \in V} d^-(x) = |E|.$$

**Exercise 1.6.6** In any digraph, we define a **walk** as a sequence

$$v_0, e_1, v_1, e_2, \ldots, e_k, v_k$$

with $v_{i-1}$ the tail of $e_i$ and $v_i$ its head. The analogous notions of trail, path, circuit, and cycle are easily extended to digraphs in the obvious way. If $X$ is a digraph such that the outdegree of every vertex is at least one, show that $X$ contains a cycle.

**Exercise 1.6.7** An **Eulerian trail** in a digraph is a trail containing all the edges. An **Eulerian circuit** is a closed trail containing all the edges. Show that a digraph $X$ contains an Eulerian circuit if and only if its underlying graph has at most one component and $d^+(v) = d^-(v)$ for every vertex $v$.

**Exercise 1.6.8** Determine for what values of $r \geq 1$ and $s \geq 1$ is the complete bipartite graph $K_{r,s}$ Eulerian.

**Exercise 1.6.9** What is the maximum number of edges in a bipartite graph with $n \geq 1$ vertices?

**Exercise 1.6.10** How many 4-cycles are in $K_{m,n}$ ?

**Exercise 1.6.11** Let $Q_n$ be the $n$-**dimensional cube** graph. Its vertices are $n$-tuples of 0 and 1 with two vertices being adjacent if they differ in precisely one position. Show that $Q_n$ is connected and bipartite.

**Exercise 1.6.12** Let $n$ be a natural number. Show that $Q_n$ has $2^n$ vertices and $n2^{n-1}$ edges.

**Exercise 1.6.13** How many 4-cycles are in $Q_n$ ?

**Exercise 1.6.14** Is $K_{2,3}$ a subgraph of $Q_n$?

**Exercise 1.6.15** Show that every graph $X$ has a bipartite subgraph with at least half the number of edges of $X$.

**Exercise 1.6.16** Let $X$ be a graph in which every vertex has an even degree. Show that it is possible to orient the edges of $X$ such that the indegree equals the outdegree for each vertex.

**Exercise 1.6.17** Show that a graph $X$ is connected if and only if, for any partition of its vertex set into two non-empty sets, there exists at least one edge between the two sets.

**Exercise 1.6.18** Show that in a connected graph, any two paths of maximum length have at least one common vertex.

**Exercise 1.6.19** Let $X$ be a graph with $n$ vertices and $m$ edges. Show that there exists at least one edge $xy$ such that

$$d(x) + d(y) \geq \frac{4m}{n}.$$

**Exercise 1.6.20** Let $n \geq 3$ be a natural number. If $X$ is a graph with $n$ vertices and $m$ edges that does not contain any triangles $K_3$ (equivalently, the clique number of $X$ is 2 or less), prove that $m \leq \lfloor \frac{n^2}{4} \rfloor$. Give an example of a graph with $n$ vertices containing no triangles with $\lfloor \frac{n^2}{4} \rfloor$ edges.

# Chapter 2
# Basic Counting

## 2.1 Finite Sets and Permutations

The earliest historical record of combinatorial questions seems to be the Chandaḥ sūtra of Piṅgala in third century BCE, India (Plofker 2009). There, we find extensive discussion about all the possible arrangements of poetic meters. There is even a discussion of the binary system. What is now called the Pascal triangle seems to have been known to the ancient Arabs, Indians, and the Chinese, several centuries before Pascal (Joseph 1991). However, modern combinatorics begins with a classic work of the French mathematician Abraham de Moivre (1667–1754) in the seventeenth century.

The main objects of study in combinatorics are finite sets. To formally define them, we need the notion of bijective function. Given two sets $A$ and $B$, a function $f : A \to B$ is **injective** or **one-to-one** if $f(a_1) \neq f(a_2)$ for any $a_1, a_2 \in A$ with $a_1 \neq a_2$. A function $f : A \to B$ is **surjective** or **onto** if, for any $b \in B$, there exist $a \in A$ such that $f(a) = b$. A function is **bijective** if it is injective and surjective. A function $f : A \to B$ is **invertible** if there exists a function $g : B \to A$ such that $f(a) = b$ if and only if $g(b) = a$. If such $g$ exists, it is not hard to see that it is unique, is called **the inverse** of $f$ and is denoted by $f^{-1}$. It is also not too hard to show that a function $f$ is bijective if and only $f$ is invertible.

We say that a set $A$ is **finite** if either $A$ is the empty set $\emptyset$ or if there exists a natural number $n$ and a bijective function $f : \{1, \ldots, n\} \to A$. If $A = \emptyset$, we say that $A$ has cardinality or size 0 while in the second situation, we say that $A$ has $n$ elements, or it has cardinality or size $n$. For a finite set $A$, we denote its cardinality by $|A|$. For every natural number $n$, we denote by $[n]$ the set $\{1, 2, \ldots, n\}$. To each subset $A$ of $[n]$, one can associate its **characteristic vector** $\chi_A \in \{0, 1\}^n$, where

$$\chi_A(a) = \begin{cases} 1 & \text{, if } a \in A \\ 0 & \text{, if } a \notin A. \end{cases}$$

© Hindustan Book Agency 2022
S. M. Cioabă and M. R. Murty, *A First Course in Graph Theory and Combinatorics*,
Texts and Readings in Mathematics 55, https://doi.org/10.1007/978-981-19-0957-3_2

**Proposition 2.1.1** *A set with $n$ elements has exactly $2^n$ subsets.*

**Proof** The result is true for $n = 0$ as the only subset of $\emptyset$ is the empty set itself. For $n \geq 1$, without loss of generality, we may assume that our set is $[n]$. The correspondence $A \mapsto \chi_A$ is a bijective function between the subsets of $[n]$ and the vectors in $\{0, 1\}^n$. The result follows from noticing that there are precisely $2^n$ vectors in $\{0, 1\}^n$.   ∎

One can also use induction on $n$ to prove the previous proposition (see Exercise 2.6.2).

A **permutation** of $[n]$ is a bijective function $f : [n] \rightarrow [n]$. A permutation can be written using a two-line notation as

$$\sigma = \begin{pmatrix} 1 & 2 & \ldots & n \\ \sigma(1) & \sigma(2) & \ldots & \sigma(n) \end{pmatrix}.$$

or in a one-line notation as

$$\sigma = \begin{pmatrix} \sigma(1) & \sigma(2) & \ldots & \sigma(n) \end{pmatrix}.$$

For a non-negative integer $n$, define $n!$ (called $n$-**factorial**) as follows:

$$n! = \begin{cases} 1 & \text{if } n = 0 \\ n \cdot (n-1)! & \text{if } n \geq 1. \end{cases}$$

**Proposition 2.1.2** *The number of permutations of $[n]$ is $n!$.*

**Proof** Let $\sigma : [n] \rightarrow [n]$ be a permutation of $[n]$. Since $\sigma(1)$ can be chosen in $n$ ways, $\sigma(2)$ in $(n-1)$ ways (as it cannot equal $\sigma(1)$), $\ldots$, $\sigma(n-1)$ in 2 ways and $\sigma(n)$ in one way, it follows that the number of permutations is $n(n-1)\ldots 2 \cdot 1 = n!$.   ∎

Given two permutations $\sigma, \tau$ of $[n]$, the function $\sigma \circ \tau : [n] \rightarrow [n]$ is defined as $(\sigma \circ \tau)(a) = \sigma(\tau(a))$ for any $a \in [n]$. The function $\tau \circ \sigma$ can be defined similarly. Both $\sigma \circ \tau$ and $\tau \circ \sigma$ are permutations of $[n]$. The following examples show that $\sigma \circ \tau \neq \tau \circ \sigma$ in general.

**Example 2.1.3** If $\sigma = \begin{pmatrix} 1 & 2 & 3 \\ 2 & 3 & 1 \end{pmatrix}$ and $\tau = \begin{pmatrix} 1 & 2 & 3 \\ 1 & 3 & 2 \end{pmatrix}$, then

$$\sigma \circ \tau = \begin{pmatrix} 1 & 2 & 3 \\ 2 & 1 & 3 \end{pmatrix}, \quad \tau \circ \sigma = \begin{pmatrix} 1 & 2 & 3 \\ 3 & 2 & 1 \end{pmatrix}.$$

We denote by $S_n$ the set of all permutations of $[n]$. The identity permutation $\varepsilon$ of order $n$ is defined as $\varepsilon(a) = a$ for any $a \in [n]$.

Together with the binary operation $\circ$ of function compositions, the set $S_n$ forms an abstract group, which means that the following properties are satisfied:

(1) Associativity: For any $\sigma, \tau, \pi \in S_n$,

$$(\sigma \circ \tau) \circ \pi = \sigma \circ (\tau \circ \pi).$$

(2) Neutral Element: For any $\sigma \in S_n$,

$$\sigma \circ \varepsilon = \varepsilon \circ \sigma = \sigma.$$

(3) Inverse: For any $\sigma \in S_n$, there exists $\tau \in S_n$ such that

$$\sigma \circ \tau = \tau \circ \sigma = \varepsilon.$$

Such $\tau$ is unique and is denoted by $\sigma^{-1}$.

The group $(S_n, \circ)$ is called the **symmetric group of degree** $n$.

We call $a \in [n]$ a **fixed point** for a permutation $\sigma$ if $\sigma(a) = a$. For $k \geq 2$ and distinct elements $a_1, \ldots, a_k \in [n]$, the **cycle** $(a_1, \ldots, a_k)$ denotes the permutation $\sigma \in S_n$ with

$$\sigma(b) = b, \ \forall b \in [n] \setminus \{a_1, \ldots, a_k\}$$

and

$$\sigma(a_1) = a_2, \ \sigma(a_2) = a_3, \ \ldots, \sigma(a_{k-1}) = \sigma(a_k), \ \sigma(a_k) = \sigma(a_1).$$

Note that $(a_1, \ldots, a_k) = (a_j, a_{j+1}, \ldots, a_k, a_1, \ldots, a_{j-1})$ for each $j \in [k]$. The **length** of the cycle $(a_1, \ldots, a_k)$ is $k$. A cycle of length two is also known as a **transposition**. Two cycles $(a_1, \ldots, a_k)$ and $(b_1, \ldots, b_\ell)$ are **disjoint** if $\{a_1, \ldots, a_k\} \cap \{b_1, \ldots, b_\ell\} = \emptyset$. We leave it as an exercise to the reader to show that if $\sigma$ and $\pi$ are disjoint cycles, then $\sigma \circ \pi = \pi \circ \sigma$.

**Theorem 2.1.4** *Every non-identity permutation can be written as a product of one or more disjoint cycles. The representation is unique modulo the order of the factors and the starting points of the cycles.*

**Proof** Let $\sigma \in S_n$ such that $\sigma \neq \varepsilon$. We prove the theorem by strong induction on the number of non-fixed points $k$ of $\sigma$.

For the base case $k = 2$, $\sigma$ must be a transposition and therefore, a cycle of length two. This proves the base case.

For the induction step, let $k \geq 3$ and assume that our statement is true for all permutations with $k - 1$ or less non-fixed points. Consider a permutation $\sigma \in S_n$ with $k$ non-fixed points. Take $a \in [n]$ such that $\sigma(a) \neq a$. For $\ell \geq 1$, denote by $\sigma^\ell$ the permutation $\underbrace{\sigma \circ \cdots \circ \sigma}_{\ell \text{ times}}$. For $\ell = 0$, $\sigma^0$ is defined as $\varepsilon$. The $n + 1$ elements $a, \sigma(a), \sigma^2(a), \ldots, \sigma^n(a)$ belong to the set $[n]$ and therefore, two of them must be equal. Hence, there exist $0 \leq s < t \leq n$ such that $\sigma^s(a) = \sigma^t(a)$ and, therefore, $\sigma^{t-s}(a) = a$. Denote by $\ell$ the smallest natural number such that $\sigma^\ell(a) = a$. Then $\pi = (a, \sigma(a), \ldots, \sigma^{l-1}(a))$ is a cycle of length $\ell$.

Consider now the permutation $\sigma' = \sigma \circ \pi^{-1}$. Any fixed point of $\sigma$ will be a fixed point of $\sigma'$. In addition, any element in $\{a, \sigma(a), \ldots, \sigma^{\ell-1}(a)\}$ will also be a fixed point of $\sigma'$. Hence, $\sigma'$ will have $k - 1$ or less non-fixed points. By the induction hypothesis, $\sigma'$ will be identity or a product of disjoint cycles. If $\sigma' = \varepsilon$, then $\sigma$ equals the cycle $(a, \sigma(a), \ldots, \sigma^{\ell-1}(a))$ and we are done. Otherwise, $\sigma'$ is a product of some disjoint cycles $\pi_1 \circ \ldots \circ \pi_t$ for some $t \geq 1$. Since $a, \sigma(a), \ldots, \sigma^{\ell-1}(a)$ are fixed points of $\sigma'$, it follows that $\pi_1, \ldots, \pi_t$ and $\pi$ are also pairwise disjoint. Hence, $\sigma = \sigma' \circ \pi = \pi_1 \circ \ldots \circ \pi_t \circ \pi$ is a decomposition of $\sigma$ into disjoint cycles.

For the uniqueness, one can use again strong induction on the number of non-fixed points. We leave the base case for the reader to sort out. For the induction step, assume that $\pi_1 \circ \ldots \circ \pi_r = \tau_1 \circ \ldots \circ \tau_s$ are two representations as products of disjoint cycles of a non-identity permutation $\sigma$. Let $a$ be a non-fixed point of $\sigma$. There must exist $j \in [r]$ and $\ell \in [s]$ such that $\pi_j(a) \neq a$ and $\tau_\ell(a) \neq a$. Because disjoint cycles commute, we may assume that $\pi_r(a) \neq a$ and $\tau_s(a) \neq a$. It follows that $\pi_r = \tau_s = (a, \sigma(a), \ldots, \sigma^{\ell-1}(a))$, where $\ell \geq 2$ is the smallest natural number such that $\sigma^\ell(a) = a$. This implies that $\pi_1 \circ \ldots \circ \pi_{r-1} = \tau_1 \circ \ldots \circ \tau_{s-1}$. Since the permutation $\pi_1 \circ \ldots \circ \pi_{r-1}$ has fewer non-fixed points than $\sigma$, by the induction hypothesis, we get that $r - 1 = s - 1$ and (modulo a renumbering of the cycles) $\pi_j = \tau_j$ for any $1 \leq j \leq r - 1$. This proves the theorem. ∎

**The order** $ord(\sigma)$ of a permutation $\sigma \in S_n$ is the smallest natural number $k$ such that $\sigma^k$ is the identity permutation. The identity permutation has order 1. The previous theorem can be used to find the order of any non-identity permutation.

**Corollary 2.1.5** *The order of a non-identity permutation in $S_n$ equals the least common multiple of the lengths of the cycles in its disjoint cycle decomposition.*

**Proof** The result can be proved using induction on the number of cycles in the decomposition, and the interested reader is invited to fill out the details. ∎

There are many ways to write a given permutation as a product of transpositions. However, one parameter will always be the same regardless of the decomposition, namely, the parity of the number of transpositions. To show this, assume that we have $n$ variables/indeterminates $X_1, \ldots, X_n$ and consider the polynomial

$$\Delta_n = \prod_{1 \leq i < j \leq n} (X_i - X_j).$$

For example, when $n = 3$,

$$\Delta_3 = (X_1 - X_2)(X_1 - X_3)(X_2 - X_3).$$

For any $\sigma \in S_n$, define the polynomial

$$\Delta_n(\sigma) = \prod_{1 \leq i < j \leq n} (X_{\sigma(i)} - X_{\sigma(j)}).$$

When $\sigma = \begin{pmatrix} 1\ 2\ 3 \\ 2\ 3\ 1 \end{pmatrix}$ and $\tau = \begin{pmatrix} 1\ 2\ 3 \\ 3\ 2\ 1 \end{pmatrix}$, we have that

$$\Delta_3(\sigma) = (X_2 - X_3)(X_2 - X_1)(X_3 - X_1) = +\Delta_3$$

and

$$\Delta_3(\tau) = (X_3 - X_2)(X_3 - X_1)(X_2 - X_1) = -\Delta_3.$$

It is not too difficult to see that for any $n \geq 2$ and $\sigma \in S_n$, $\Delta_n(\sigma)$ equals $+\Delta_n$ or $-\Delta_n$. **The signature** $\mathrm{sgn}(\sigma)$ of a permutation $\sigma$ is defined as follows:

$$\mathrm{sgn}(\sigma) = \begin{cases} +1 & \text{if } \Delta_n(\sigma) = +\Delta_n \\ -1 & \text{if } \Delta_n(\sigma) = -\Delta_n. \end{cases}$$

A permutation $\sigma \in S_n$ is called **even** if $\mathrm{sgn}(\sigma) = +1$ and is called **odd** if $\mathrm{sgn}(\sigma) = -1$. The identity permutation is even.

**Proposition 2.1.6** *Let $n \geq 2$.*

*(1) If $\sigma$ is a transposition, then $\Delta_n(\sigma) = -\Delta_n$.*
*(2) For any $\sigma, \tau \in S_n$, $\mathrm{sgn}(\sigma \circ \tau) = \mathrm{sgn}(\sigma)\mathrm{sgn}(\tau)$.*
*(3) If $\sigma$ is the product of $k$ transpositions, then $k$ is even if and only if $\mathrm{sgn}(\sigma) = +1$.*

**Proof** For the first part, assume that $\sigma = (a, b)$ for some $1 \leq a < b \leq n$. We analyze the difference in sign between the factors of $\Delta_n$ and $\Delta_n(\sigma)$. Any factor of the form $X_i - X_j$, where $i < j$ and $\{i, j\} \cap \{a, b\} = \emptyset$ stays the same in both $\Delta_n$ and $\Delta_n(\sigma)$. When $i < a$, the factors $X_i - X_a$ and $X_i - X_b$ in $\Delta_n$ become $X_i - X_b$ and $X_i - X_a$, respectively, in $\Delta_n(\sigma)$ and their product stays the same. When $j > b$, the factors $X_a - X_j$ and $X_b - X_j$ in $\Delta_n$ become $X_b - X_j$ and $X_a - X_j$ in $\Delta_n$, respectively. Again, the product of these factors stays the same. When $a < i < b$, $X_a - X_i$ and $X_i - X_b$ are replaced by $X_b - X_i$ and $X_i - X_a$, respectively. The product of these factors is the same in both $\Delta_n$ and $\Delta_n(\sigma)$. Finally, the factor $X_a - X_b$ in $\Delta_n$ changes sign and becomes $X_b - X_a$ in $\Delta_n(\sigma)$. Hence, $\Delta_n(\sigma) = -\Delta_n$.

For the second part, let $\sigma, \tau \in S_n$. Assume that $\Delta_n(\tau)$ has exactly $r$ factors of the form $X_j - X_i$ with $j > i$ and thus, $\Delta_n(\tau) = (-1)^r \Delta_n$. Therefore, $\Delta_n(\sigma \circ \tau)$ has exactly $r$ factors of the form $X_{\sigma(j)} - X_{\sigma(i)}$ with $j > i$. Changing the signs in all these $r$ factors, we get that

$$\Delta_n(\sigma \circ \tau) = (-1)^r \prod_{1 \leq a < b \leq n} (X_{\sigma(a)} - X_{\sigma(b)}) = \mathrm{sgn}(\tau)\mathrm{sgn}(\sigma)\Delta_n.$$

This implies the second assertion. We leave the proof of the third part as an exercise for the motivated reader. ∎

Using this proposition, one can deduce that the product of any two even permutations is also even. The set of all even permutations in $S_n$ forms a group called **the alternating group of degree** $n$ and is denoted by $A_n$.

For any integer $k$ with $0 \le k \le n$, define the **binomial coefficient** $\binom{n}{k}$ as the number of subsets with $k$ elements (or $k$-subsets) of $[n]$.

**Proposition 2.1.7** *For any integers $n \ge k \ge 0$,*

$$\binom{n}{k} = \frac{n(n-1)\ldots(n-k+1)}{k!} = \frac{n!}{k!(n-k)!}.$$

**Proof** If $n = k = 0$, then $\binom{0}{0} = 0$. If $n \ge 1$, then it is not too difficult to see that $\binom{n}{0} = 1$ and $\binom{n}{1} = n$. All these formulas agree with the equations above.

For $n \ge k \ge 2$, let us count the number of pairs $(A, x)$, where $A \subseteq [n]$, $|A| = k$ and $x \in A$. There are $\binom{n}{k}$ such $A$'s and each has $k$ elements. Thus, the answer is $k\binom{n}{k}$. On the other hand, if we count the $x$'s first, we have $n$ choices. For each $x$, there are $\binom{n-1}{k-1}$ subsets $A$ such that $A \subseteq [n]$, $x \in A$. This is because each such $A$ is of the form $B \cup \{x\}$, where $B \subset [n] \setminus \{x\}$ and $|B| = k - 1$. Thus, the answer we get now is $n\binom{n-1}{k-1}$. Hence, $\binom{n}{k} = \frac{n}{k}\binom{n-1}{k-1}$.

Replacing $n$ by $n-1, n-2, \ldots, n-k+2$, we obtain

$$\binom{n-j}{k-j} = \frac{n-j}{k-j}\binom{n-j-1}{k-j-1}$$

for $j = 0, 1, \ldots, k-2$. Multiplying all these equations together, we get

$$\binom{n}{k}\prod_{j=1}^{k-2}\binom{n-j}{k-j} = \frac{n(n-1)\ldots(n-k+1)}{k!}\prod_{j=1}^{k-2}\binom{n-j}{k-j}.$$

Simplifying the previous equality, we obtain

$$\binom{n}{k} = \frac{n(n-1)\ldots(n-k+1)}{k!} = \frac{\frac{n!}{(n-k)!}}{k!} = \frac{n!}{k!(n-k)!}.$$

This finishes our proof.  ∎

One can use Proposition 2.1.7 to show that the middle coefficient $\binom{n}{n/2}$ when $n$ is even or the middle coefficients $\binom{n}{\frac{n-1}{2}} = \binom{n}{\frac{n+1}{2}}$ are the largest among all binomial coefficients $\binom{n}{k}$ for $0 \le k \le n$ (see Exercise 2.6.1).

The formulae for the number of permutations with $n$ elements and the number of $k$-subsets of $[n]$ were known to Bhaskara around 1150. Special cases of these formulae were found in texts dating back to the second century BC. The following theorem is often attributed to Blaise Pascal (1623–1662) as it appeared in a posthumous pamphlet published in 1665. The result was known to various mathematicians preceding Pascal, such as the third-century Indian mathematician Pingala.

**Theorem 2.1.8** (Binomial Theorem) *For any natural number $n$ and any real numbers $x$ and $y$,*

$$(x + y)^n = \sum_{k=0}^{n} \binom{n}{k} x^k y^{n-k}.$$

**Proof** Writing $(x + y)^n = \underbrace{(x + y)(x + y) \ldots (x + y)}_{n \text{ times}}$, we notice that the number of times the term $x^k y^{n-k}$ appears in the expansion of $(x + y)^n$, equals the number of ways of choosing exactly $k$ brackets (for $x$) from the $n$ factors of the product. That is exactly $\binom{n}{k}$. Since $k$ can take any value between 0 and $n$, this gives the desired result. ∎

Sir Isaac Newton (1643–1727) was one of the greatest mathematicians in the history of the world. His contributions to mathematics, physics, and astronomy are deep and numerous. In 1676, Newton showed that a similar formula holds for real $n$. Newton's formula involves infinite series, and it will be discussed in the Catalan number section.

If $f, g : \mathbb{N} \to \mathbb{R}$, we say $f(n) \sim g(n)$ if $\lim_{n \to \infty} \dfrac{f(n)}{g(n)} = 1$. James Stirling (1692–1770) was a Scottish mathematician who showed that

$$n! \sim \sqrt{2\pi n} \left(\frac{n}{e}\right)^n.$$

This is usually called Stirling's formula. It appears in *Methodus Differentialis* which Stirling published in 1730. Abraham de Moivre (1667–1754) also knew such a result around 1730 but did not know the precise constant of $\sqrt{2\pi}$.

## 2.2 Fibonacci Numbers

The term *recurrence* is due to Abraham de Moivre (1722). In many counting questions, it is more expedient to obtain a recurrence relation for the combinatorial quantity in question. Depending on the nature of this recurrence, one is then able to determine in some cases, an explicit formula, and in other cases, where explicit formulas are lacking, some idea of the growth of the function. We will give several examples in this chapter.

Leonardo Pisano Fibonacci (1170–1250) was an Italian mathematician whose 1202 book *Liber Abaci—The Book of the Abacus* introduced the use of the Hindu-Arabic numerals into Europe. This is the decimal system that we now use worldwide. In the same book, he studied the famous rabbit problem:

*A person has a pair of rabbits. Every month, each pair produces a new pair of rabbits that can also reproduce after two months. How many pairs of rabbits will be there after one year?*

The **Fibonacci numbers** $(F_n)_{n\geq 1}$ are defined recursively as follows:

$$F_1 = 1,\ F_2 = 1,\ \text{and}\ F_n = F_{n-1} + F_{n-2},\ \text{for}\ n \geq 2.$$

It turns out that $F_n$ equals the number of pairs of rabbits after $n$ months. In the table below, we list the first twelve Fibonacci numbers.

| $n$ | 1 | 2 | 3 | 4 | 5 | 6 | 7 | 8 | 9 | 10 | 11 | 12 |
|---|---|---|---|---|---|---|---|---|---|---|---|---|
| $F_n$ | 1 | 1 | 2 | 3 | 5 | 8 | 13 | 21 | 34 | 55 | 89 | 144 |

The Fibonacci numbers satisfy a **linear recurrence relation with constant coefficients**. We will describe the more general formulation later, but in this case, the characteristic equation of this recurrence relation is

$$x^2 = x + 1.$$

This has two distinct real roots: $\alpha = \frac{1+\sqrt{5}}{2}$ and $\beta = \frac{1-\sqrt{5}}{2}$. The key idea is that each of the sequences $(\alpha^n)_{n\geq 1}$ and $(\beta^n)_{n\geq 1}$ satisfies the recurrence relation above. Thus, it makes sense to make a guess that $F_n = c\alpha^n + d\beta^n$, where $c$ and $d$ are constants to be determined. Since $1 = F_1 = c\alpha + d\beta$ and $1 = F_2 = c\alpha^2 + d\beta^2$, we obtain that $1 = c\alpha^2 + d\beta^2 = c(\alpha + 1) + d(\beta + 1) = (c\alpha + d\beta) + (c + d)$ implying that $c + d = 0$. Using $c\alpha + d\beta = 1$, one gets that $c = \frac{1}{\sqrt{5}}$ and $d = \frac{-1}{\sqrt{5}}$. Hence,

$$F_n = \frac{1}{\sqrt{5}}\left[\left(\frac{1+\sqrt{5}}{2}\right)^n - \left(\frac{1-\sqrt{5}}{2}\right)^n\right]. \tag{2.2.1}$$

A short calculation gives us that $\alpha = \frac{1+\sqrt{5}}{2} \approx 1.618$ and $\beta = \frac{1-\sqrt{5}}{2} \approx -0.618$. Hence, the term $\beta^n$ will be negligible and one can show that

$$F_n = \left\lfloor \frac{\alpha^n + 1/2}{\sqrt{5}} \right\rfloor, \tag{2.2.2}$$

where $\lfloor x \rfloor$ denotes the largest integer less than or equal to $x$. When $n = 12$, we have that $\left\lfloor \frac{\alpha^{12}+1/2}{\sqrt{5}} \right\rfloor \approx \lfloor 144.54 \rfloor = 144$ which confirms our computations from the table above.

The same method can be used for more general linear recurrence relations such as

$$y_n = a_1 y_{n-1} + a_2 y_{n-2} + \cdots + a_k y_{n-k},$$

where $k \geq 1$ is a fixed integer and $a_1, a_2, \ldots, a_k$ are all constant (they do not depend on $n$).

To find a general formula for $y_n$, one first solves the characteristic equation

$$x^k = a_1 x^{k-1} + a_2 x^{k-2} + \cdots + a_k.$$

If this equation has $k$ distinct solutions, then $y_n$ is going to be a linear combination of the $n$th powers of these solutions. Using the initial $k$ values of the sequence $(y_n)_n$, one can find the exact formula for $y_n$. If the equation above has multiple solutions, a formula for $y_n$ can be determined as follows. If $\alpha$ is a solution with multiplicity $r$, then one can check $\alpha^n, n\alpha^n, \ldots, n^{r-1}\alpha^n$ are all solutions of the characteristic equation. We can write $y_n$ as a linear combination of such solutions and use the initial values of the sequence $(y_n)_n$ to determine a precise formula.

The previous results can be interpreted using the language of formal power series. To an infinite sequence $(a_n)_{n\geq 0}$, we associate the following **formal power series** also called **the ordinary generating series** of the sequence:

$$\sum_{n\geq 0} a_n t^n.$$

We regard such series as algebraic objects without any interest in their convergence. We say two series are equal if their coefficient sequences are identical. We define addition and subtraction as follows:

$$\sum_{n\geq 0}(a_n \pm b_n)t^n = \sum_{n\geq 0} a_n t^n \pm \sum_{n\geq 0} b_n t^n.$$

The multiplication is defined similarly to the one for polynomials.

$$\sum_{n\geq 0} a_n t^n \cdot \sum_{n\geq 0} b_n t^n = \sum_{n\geq 0} c_n t^n,$$

where $c_n = \sum_{k=0}^{n} a_k b_{n-k}$. This implies that

$$\frac{1}{1-t} = \sum_{n\geq 0} t^n, \tag{2.2.3}$$

because $(1-t)\sum_{n\geq 0} t^n = 1$. We can also differentiate formal power series the same way as one would do for polynomials.

$$\left(\sum_{n\geq 0} a_n t^n\right)' = \sum_{n\geq 1} n a_n t^{n-1}.$$

The standard functions of analysis are defined as formal power series by their usual Taylor series. For example,

$$e^t = \sum_{n\geq 0} \frac{t^n}{n!}.$$

The following equation is a definition of $(1 + t)^\alpha$

$$(1 + t)^\alpha = \sum_{n \geq 0} \binom{\alpha}{n} t^n, \qquad (2.2.4)$$

where $\binom{\alpha}{n} = \frac{\alpha(\alpha-1)\ldots(\alpha-n+1)}{n!}$ for any real number $\alpha$. If $\alpha$ is a non-negative integer, then this is just Theorem 2.1.8 since $\binom{\alpha}{n} = 0$ for $n > \alpha$. For $\alpha$ real, the equation above will be regarded here as a definition. An alternative approach would be to define $(1 + t)^\alpha$ for any rational $\alpha$ by using the exponent laws (which hold for power series) and then prove that its Taylor series has the claimed form. This was done by Newton.

The generating function for the Fibonacci numbers is denoted by $f(t) = \sum_{n \geq 0} F_n t^n$, where we define $F_0 = 0$. It follows that

$$f(t) = \sum_{n \geq 0} F_n t^n = F_1 t + \sum_{n \geq 2} F_n t^n = t + \sum_{n \geq 2}(F_{n-1} + F_{n-2})t^n$$
$$= t + \sum_{n \geq 2} F_{n-1} t^n + \sum_{n \geq 2} F_{n-2} t^n$$
$$= t + tf(t) + t^2 f(t).$$

Hence, $f(t) = \frac{t}{1-t-t^2} = \frac{t}{(1-t\alpha)(1-t\beta)} = \frac{1/\sqrt{5}}{1-t\alpha} - \frac{1/\sqrt{5}}{1-t\beta}$. From Eq. (2.2.3), we get that $\frac{1}{1-t\alpha} = \sum_{n \geq 0}(t\alpha)^n$ and $\frac{1}{1-t\beta} = \sum_{n \geq 0}(t\beta)^n$. Therefore,

$$\sum_{n \geq 0} F_n t^n = f(t) = \sum_{n \geq 0} \frac{1}{\sqrt{5}}(\alpha^n - \beta^n)t^n,$$

implying again that $F_n = \frac{1}{\sqrt{5}}(\alpha^n - \beta^n)$.

## 2.3  Catalan Numbers

Eugéne Charles Catalan (1814–1894) was born in Bruges, Belgium. The numbers that bear his name were actually studied by Leonhard Euler (1707–1783) in 1751 as follows. For $n \geq 1$, $C_n$ equals the number of ways of decomposing a convex $n + 2$-gon into triangles by $n - 1$ non-intersecting diagonals. We set $C_0 = 1$ and we have that $C_1 = 1$, $C_2 = 2$, and $C_3 = 5$ (see Fig. 2.1). In his study of these numbers, Euler corresponded with the German mathematician Christian Goldbach (1690–1764) and the Hungarian-German mathematician János András Segner (1704–1777). In the older literature, these numbers are referred to as Euler numbers or Euler-Segner numbers. While Catalan also studied these numbers in 1838, the name *Catalan*

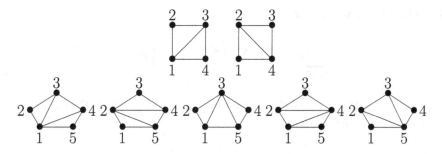

**Fig. 2.1** $C_2 = 2$ and $C_3 = 5$.

*numbers* seems to have been popularized by the American mathematicians John Riordan (1903–1988) in 1968 and Martin Gardner (1914–2010) in 1976.

The Catalan numbers have many combinatorial interpretations and arise in branches of mathematics and computer science. There are 214 interpretations of the Catalan numbers in Richard Stanley's book *Catalan Numbers*. The book also contains an appendix by Igor Pak describing the fascinating history of these numbers and the people who have studied them throughout the years.

We first obtain a recurrence for $C_n$ as follows. Consider a triangulation of a $(n + 3)$-gon $P$ and fix an edge of it, say the one connecting 1 and $n + 3$. This edge is contained in exactly one triangle of our triangulation, and removing it will result in a triangulated $(k + 2)$-gon $P_1$ and a triangulated $(n - k + 2)$-gon $P_2$. This procedure is reversible, and any triangulation of $P_1$ can be paired up with any triangulation $P_2$ to give a unique triangulation of $P$. Since there are $C_k$ choices for $P_1$ and $C_{n-k}$ choices for $P_2$, so we get that

$$C_{n+1} = \sum_{k=0}^{n} C_k C_{n-k}. \tag{2.3.1}$$

It seems that Segner was the first to notice this recurrence relation which we can use to calculate some other Catalan numbers in the table below.

| $n$ | 0 | 1 | 2 | 3 | 4 | 5 | 6 | 7 | 8 |
|---|---|---|---|---|---|---|---|---|---|
| $C_n$ | 1 | 1 | 2 | 5 | 14 | 42 | 132 | 429 | 1430 |

Notice that the recurrence relation (2.3.1) is more complicated than the one in the previous section. In order to determine a nice formula for the Catalan numbers, we use the theory of generating functions. We encode the Catalan numbers in its generating function as follows:

$$F(t) = \sum_{n \geq 0} C_n t^n.$$

From Eq. (2.3.1), we get that

$$\sum_{n \geq 0} C_{n+1} t^n = \sum_{n \geq 0} \left( \sum_{k=0}^{n} C_k C_{n-k} \right) t^n.$$

The left-hand side $\sum_{n \geq 0} C_{n+1} t^n$ also equals $\sum_{n \geq 1} C_n t^{n-1} = \frac{F(t)-1}{t}$. The right-hand side above equals $F(t)^2$. Hence,

$$t F(t)^2 - F(t) + 1 = 0,$$

which implies that $F(t) = \frac{1 \pm \sqrt{1-4t}}{2t}$. From Eq. (2.2.4), we get that

$$\sqrt{1 - 4t} = \sum_{n \geq 0} \binom{1/2}{n} (-4t)^n = 1 - 2t - 2t^2 - 4t^3 - 10t^4 - \dots.$$

If $F(t) = \frac{1 + \sqrt{1-4t}}{2t} = 1/t - 1 - t - 2t^2 - 5t^3 - \dots$, then we get a contradiction as $F(t)$ contains only non-negative powers of $t$. Hence, it must be that $F(t) = \frac{1 - \sqrt{1-4t}}{2t}$ and therefore,

$$F(t) = \frac{1 - \sum_{n \geq 0} \binom{1/2}{n}(-4t)^n}{t} = \frac{1 - 1 - \sum_{n \geq 1} \binom{1/2}{n}(-4)^n t^n}{2t}$$

$$= \sum_{n \geq 1} \frac{-\binom{1/2}{n}(-4)^n}{2} t^{n-1} = \sum_{n \geq 0} \frac{-\binom{1/2}{n+1}(-4)^{n+1}}{2} t^n.$$

Hence, $C_n = \frac{-\binom{1/2}{n+1}(-4)^{n+1}}{2}$ for each $n \geq 0$. This is perhaps not the nicest formula around, but it can be simplified as follows.

**Theorem 2.3.1** *For $n \geq 1$,*

$$C_n = \frac{\binom{2n}{n}}{n+1}. \tag{2.3.2}$$

***Proof*** Note that

$$\binom{1/2}{n+1} = \frac{\prod_{k=0}^{n} \left( \frac{1}{2} - k \right)}{(n+1)!} = (-1)^n \frac{\prod_{k=1}^{n}(2k-1)}{2^{n+1}(n+1)!} = (-1)^n \frac{\frac{(2n)!}{2^n n!}}{2^{n+1}(n+1)!}$$

$$= (-1)^n \frac{(2n)!}{2^{2n+1} n!(n+1)!}.$$

Therefore,

$$C_n = \frac{-\binom{1/2}{n+1}(-4)^{n+1}}{2} = \frac{(-1)^{n+1}\frac{(2n)!}{2^{2n+1}n!(n+1)!} \cdot (-4)^{n+1}}{2}$$

$$= \frac{(2n)!}{n!(n+1)!} = \frac{\binom{2n}{n}}{n+1},$$

as claimed. ∎

We can use Stirling's formula to determine the asymptotic behaviour of $C_{n+1}$. Indeed, by Stirling's formula,

$$n! \sim \sqrt{2\pi n}(n/e)^n,$$

so that

$$C_n \sim \frac{2^{2n}}{(n+1)\sqrt{\pi n}},$$

from which we see that the Catalan numbers have exponential growth.

We give now another combinatorial interpretation of $C_n$ as the number of ways we can bracket a sum of $n+1$ elements so that it can be calculated by adding two terms at a time. For example, for $n = 2$, we have

$$((x_1 + x_2) + x_3) \quad \text{and} \quad (x_1 + (x_2 + x_3)).$$

Thus, $C_2 = 2$.

For $n = 3$, we have that $C_3 = 5$ since there are five ways of bracketing a sum with 4 terms:

$$(((x_1 + x_2) + x_3) + x_4),$$

$$(x_1 + ((x_2 + x_3) + x_4)),$$

$$((x_1 + x_2) + (x_3 + x_4)),$$

$$((x_1 + (x_2 + x_3)) + x_4),$$

$$(x_1 + (x_2 + (x_3 + x_4))).$$

In the triangulations of a 5-gon in Fig. 2.1, let us label the non-horizontal sides as follows: 12 by $x_1$, 23 by $x_2$, 34 by $x_3$, and 45 by $x_4$. One may be able to figure a general bijective correspondence between the triangulations of a $(n + 2)$-gon and the bracketings of a sum with $n + 1$ terms by noting that the first triangulation from the left in Fig. 2.1 may be associated with $(((x_1 + x_2) + x_3) + x_4)$ and the second triangulation from the left can be related to $(x_1 + ((x_2 + x_3) + x_4))$.

## 2.4  Derangements and Involutions

Consider the problem of counting the number of permutations $\sigma$ of $S_n$ without any fixed points. These are permutations with the property that $\sigma(i) \neq i$ for all $1 \leq i \leq n$. Such permutations are called **derangements**. The first appearance of this problem is in 1708 in a book on games of chance *Essay d'Analyse sur les Jeux de Hazard-Analysis of Games of Chance* by the French mathematician Pierre Raymond de Montmort (1678–1719).

Let $d_n$ be the number of derangements on $[n]$. For small values of $n$, one can figure out some values:

| $n$ | 1 | 2 | 3 | 4 |
|-----|---|---|---|---|
| $d_n$ | 0 | 1 | 2 | 9 |

but perhaps a closed formula is not so clear. We will obtain a recurrence relation for $d_n$ as follows. For such a derangement $\sigma \in S_n$, we know that $\sigma(n) = i$ for some $1 \leq i \leq n - 1$. We fix such an $i$ and count the number of derangements with $\sigma(n) = i$. Since there are $n - 1$ choices for $i$, the final tally is obtained by multiplying this number by $n - 1$. If $\sigma$ is a derangement with $\sigma(n) = i$, we consider two cases. If $\sigma(i) = n$, then $\sigma$ restricted to $\{1, 2, ..., n\}\setminus\{i, n\}$ is a derangement on $n - 2$ letters and the number of such is $d_{n-2}$. If $\sigma(i) \neq n$, let $j$ be such that $\sigma(j) = n$, with $i \neq j$. We define $\sigma'$ by setting

$$\sigma'(k) = \sigma(k), \quad \text{for } 1 \leq k \leq n - 1, \ k \neq j$$

, and $\sigma'(j) = i$. Note that $\sigma'$ is a derangement on $n - 1$ letters. Conversely, if $\sigma'$ is a derangement on $n - 1$ letters and $\sigma'(j) = i$, we can extend it to a derangement on $n$ letters by setting $\sigma(j) = n$ and $\sigma(n) = i$. Thus, we proved the following result.

**Theorem 2.4.1** *For $n \geq 2$,*

$$d_n = (n - 1)(d_{n-1} + d_{n-2}). \tag{2.4.1}$$

This formula may be useful to expand the table of small values above, but it is not so obvious how to use it to get a nice formula for $d_n$. To achieve this goal, we use generating functions of a different sort. Define **the exponential generating function** of the sequence $(d_n)_{n\geq0}$ as:

$$g(t) = \sum_{n\geq0} \frac{d_n}{n!} t^n.$$

For an explanation on when to use ordinary generating functions vs exponential generating functions, the reader is invited to consult the freely available book *generatingfunctionology* by the American mathematician Herbert S. Wilf (1931–2012). In our case, we rewrite Eq. (2.4.1) as $d_{n+1} = nd_n + nd_{n-1}$. After multiplying each side by $t^n/n!$ and adding both sides up, we get that

$$\sum_{n\geq 0} \frac{d_{n+1}t^n}{n!} = \sum_{n\geq 1} \frac{nd_n t^n}{n!} + \sum_{n\geq 1} \frac{nd_{n-1}t^n}{n!}.$$

It is not too hard to see that

$$g'(t) = \sum_{n\geq 0} \frac{d_{n+1}t^n}{n!}$$

$$tg'(t) = \sum_{n\geq 1} \frac{nd_n t^n}{n!}$$

$$tg(t) = \sum_{n\geq 1} \frac{nd_{n-1}t^n}{n!}.$$

Putting all these together, we get that $g'(t) = tg'(t) + tg(t)$ which implies that

$$g(t) = \frac{e^{-t}}{1-t}.$$

The following result will be a consequence of our work.

**Theorem 2.4.2** *For $n \geq 1$,*

$$d_n = n! \sum_{j=0}^{n} \frac{(-1)^j}{j!}.$$

**Proof** As seen earlier, $\frac{1}{1-t} = \sum_{n\geq 0} t^n$ and $e^{-t} = \sum_{n\geq 0} \frac{(-t)^n}{n!}$. The result follows by using the product rule for formal power series. ∎

Let us observe that

$$\lim_{n\to\infty} \frac{d_n}{n!} = e^{-1}.$$

In fact, we can make this more precise. If $(a_n)_{n\geq 0}$ is a decreasing sequence of positive real numbers tending to zero, then one can show that the series $\sum_{j\geq 0}(-1)^j a_j$ converges and

$$\left| \sum_{j\geq 0}(-1)^j a_j - \sum_{j=0}^{n}(-1)^j a_j \right| \leq a_{n+1}.$$

It follows that

$$\left| e^{-1} - \frac{d_n}{n!} \right| < \frac{1}{(n+1)!}.$$

This implies the following result.

**Theorem 2.4.3** *For $n \geq 1$,*

$$d_n = \lfloor n!/e + 1/2 \rfloor.$$

***Proof*** By our remarks above,

$$|d_n - n!/e| \le \frac{1}{n+1},$$

for $n \ge 1$. This means that $d_n$ is uniquely determined by this inequality as the nearest integer to $n!/e$. We leave as an easy exercise for the reader to show that the nearest integer to $x$ is $\lfloor x + 1/2 \rfloor$.                                                      ∎

An interesting consequence of this result is that if $n$ people check their umbrellas at the theater at the beginning of the show, then the probability that none of them receives their own umbrella at the end of the show is $e^{-1} \approx 0.367$. So it would be wise to bet that at least one person gets his or her own umbrella at the end of the show.

Let us now count the number of elements of order at most two in the symmetric group $S_n$. Such an element is called an **involution**. Let $s(n)$ be the number of involutions in $S_n$. These numbers are also called telephone numbers as they count the number of ways of connecting $n$ telephones lines so that one line can only be connected to at most one other line. For small $n$, we get the following:

| $n$    | 1 | 2 | 3 | 4  |
|--------|---|---|---|----|
| $s(n)$ | 1 | 2 | 4 | 10 |

Recall that any non-identity permutation is a product of disjoint cycles, and the order of the permutation is the least common multiple of the cycle lengths. Thus, if the permutation has order two, then all the cycles in its decomposition must be of length two. We partition these involutions into two groups: those that fix $n$ and those that do not. The number fixing $n$ is clearly $s(n-1)$. If $\sigma$ is an involution not fixing $n$, then $\sigma(n) = i$ (say) for some $1 \le i \le n-1$. But then we must necessarily have $\sigma(i) = n$ as $\sigma$ is a product of 1-cycles or 2-cycles (transpositions). Thus, $\sigma$ restricted to $\{1, 2, \ldots, n-1\} \setminus \{i\}$ is an involution on $n-1$ letters. There are $s(n-2)$ such elements and $n-1$ choices for $i$, so we get following result.

**Theorem 2.4.4** *Let $s(n)$ be the number of involutions in $S_n$. Then*

$$s(n) = s(n-1) + (n-1)s(n-2).$$

One may use this result to figure out the exponential generating function $h(t) = \sum_{n \ge 0} \frac{s(n)t^n}{n!}$ by proving that $h'(t) = (1+t)h(t)$ and $h(t) = e^{t + \frac{t^2}{2}}$. This does not give a simple closed formula for $s(n)$, but it may lead to figuring out the asymptotics of $s(n)$ as

$$s(n) \approx \frac{n^{n/2}}{\sqrt{2}e^{n/2 + 1/4 - \sqrt{n}}}.$$

We do not include the details, but we conclude this section with another formula for $s(n)$. For a natural number $k$, denote by $(2k-1)!!$ the product $(2k-1) \cdot \ldots \cdot 1$, where each factor is odd.

**Theorem 2.4.5** *For $n \geq 2$,*

$$s(n) = \sum_{k=0}^{\lfloor \frac{n}{2} \rfloor} \binom{n}{2k} (2k-1)!!.$$

*Proof* **A matching** in a graph is a collection of pairwise disjoint edges. An involution in $S_n$ with $k$ disjoint cycles is a matching of $k$ edges in $K_n$. Since the number of matchings of order $k$ in the complete graph $K_{2k}$ equals $(2k-1) \cdot \ldots \cdot 1 = (2k-1)!!$, this gives the desired formula. ∎

## 2.5 Bell Numbers

Eric Temple Bell (1883–1960) was born in Aberdeen, Scotland. He was the president of the Mathematical Association of America between 1931 and 1933. Bell was also a prolific writer of science fiction under the name John Taine.

**The $n$th Bell number**, denoted by $B_n$, is the number of partitions of an $n$-element set. By convention, $B_0 = 1$. Clearly, $B_1 = 1$, $B_2 = 2$, and $B_3 = 5$. We derive a recurrence relation for the $B_n$. Of the partitions of an $n$-element set, we consider the block (or partition) to which $n$ belongs. Clearly, such a block can be written as $\{n\} \cup Y$ for some subset $Y$ of $\{1, 2, \ldots, n-1\}$. If this block has $k$ elements, then $Y$ is a subset of $k-1$ elements. The number of ways of choosing $Y$ is $\binom{n-1}{k-1}$. The remaining elements can be partitioned in $B_{n-k}$ ways. Thus, we obtain the following result.

**Theorem 2.5.1** *For $n \geq 1$,*

$$B_n = \sum_{k=1}^{n} \binom{n-1}{k-1} B_{n-k}.$$

We can use this recurrence to determine the exponential generating function:

$$G(t) = \sum_{n \geq 0} \frac{B_n}{n!} t^n.$$

Then,

$$G'(t) = \sum_{n \geq 1} \frac{B_n}{(n-1)!} t^{n-1} = \sum_{n \geq 1} \sum_{k=1}^{n} \frac{t^{k-1}}{(k-1)!} \cdot \frac{B_{n-k} t^{n-k}}{(n-k)!} = e^t G(t).$$

Thus, $G(t) = A e^{e^t}$ for some constant $A$. Since $G(0) = 1$, we must have $A = e^{-1}$. Hence,

$$\sum_{n \geq 0} \frac{B_n t^n}{n!} = e^{e^t - 1} = \frac{1}{e} \sum_{j \geq 0} \sum_{n \geq 0} \frac{j^n t^n}{n! j!},$$

and, on comparing the coefficients of $t^n$, we obtain the following result.

**Theorem 2.5.2** *For $n \geq 1$,*

$$B_n = \frac{1}{e} \sum_{j \geq 0} \frac{j^n}{j!}.$$

## 2.6  Exercises

**Exercise 2.6.1**  For $0 \leq k \leq n$, show that $\binom{n}{k} \leq \binom{n}{k+1}$ if and only if $k \leq \lfloor \frac{n}{2} \rfloor$.

**Exercise 2.6.2**  Show that for any natural numbers $n \geq k$,

$$\left( \frac{n}{k} \right)^k \leq \binom{n}{k} < \left( \frac{en}{k} \right)^k.$$

**Exercise 2.6.3**  Let $n$ be a natural number and $k$ an integer between 0 and $n$. If $k < n$, show that

$$\binom{n}{0} - \binom{n}{1} + \cdots + (-1)^k \binom{n}{k}$$

is positive if $k$ is even and is negative if $k$ is odd. If $k = n$, prove that

$$\binom{n}{0} - \binom{n}{1} + \cdots + (-1)^n \binom{n}{n} = 0.$$

**Exercise 2.6.4**  For each $n \geq k \geq \ell \geq 0$, show that

$$\binom{n}{k} \binom{k}{\ell} = \binom{n}{\ell} \binom{n - \ell}{k - \ell}.$$

**Exercise 2.6.5**  For $n \geq k \geq 0$,

$$\binom{n}{k} = \binom{n - 1}{k} + \binom{n - 1}{k - 1}.$$

**Exercise 2.6.6**  For any non-negative integers $n$ and $k$, show that

$$\binom{n + k + 1}{k} = \sum_{j=0}^{k} \binom{n + j}{j}.$$

**Exercise 2.6.7** For any non-negative integers $m, n$ and $k$, show that

$$\binom{m+n}{k} = \sum_{i=0}^{k} \binom{m}{i}\binom{n}{k-i}.$$

**Exercise 2.6.8** Let $n$ be a natural number. Prove that

$$\frac{2^{2n}}{2n+1} < \binom{2n}{n} < 2^{2n}.$$

Use Stirling's formula to show that

$$\binom{2n}{n} \sim \frac{2^{2n}}{\sqrt{\pi n}}.$$

**Exercise 2.6.9** Give a solution using binomial coefficients and a direct combinatorial solution to the following question: How many pairs $(A, B)$ of subsets of $[n]$ are there such that $A \cap B = \emptyset$ ?

**Exercise 2.6.10** Let $n$ be a natural number. Show that the number of even subsets of $[n]$ equals the number of odd subsets of $[n]$. Give two proofs, one using binomial formula, and one using a direct bijection. What are the sums of the sizes of these subsets?

**Exercise 2.6.11** Let $n$ be a natural number. For $r \in \{0, 1, 2\}$, let $s_r$ denote the number of subsets of $[n]$ whose order is congruent to $r$ (mod 3) for $r \in \{0, 1, 2\}$. Determine $s_0, s_1, s_2$ in terms of $n$.

**Exercise 2.6.12** If $F_n$ denotes the $n$th Fibonacci number, show that

$$\begin{bmatrix} F_{n+1} \\ F_n \end{bmatrix} = \begin{bmatrix} 1 & 1 \\ 1 & 0 \end{bmatrix} \begin{bmatrix} F_n \\ F_{n-1} \end{bmatrix}.$$

Use this relation and the eigenvalues of the above matrix to give another proof of (2.2.1).

**Exercise 2.6.13** Let $n$ be a natural number. Determine the number of solutions $(x_1, \ldots, x_{2n})$ of the equation

$$x_1 + \cdots + x_{2n} = 0$$

such that $x_1, \ldots, x_{2n} \in \{+1, -1\}$ and $\sum_{j=1}^{k} x_j \geq 0$ for each $1 \leq k \leq 2n$.

**Exercise 2.6.14** Let $n \geq k$ be two natural numbers. Consider the

$$x_1 + \cdots + x_k = n$$

Show that there $\binom{n+k-1}{k-1}$ non-negative integer solutions and there are $\binom{n-1}{k-1}$ natural number solutions of the equation above.

**Exercise 2.6.15** Let $n$ be a natural number. Prove that

$$F_n F_{n+1} = \sum_{k=1}^{n} F_k^2,$$

where $F_n$ denotes the $n$th Fibonacci number. This result has a nice geometric interpretation. Can you find it?

**Exercise 2.6.16** Let $n$ be a natural number. Calculate

$$\sum_{A \subseteq [n]} |A|^2.$$

**Exercise 2.6.17** Let $t$ be a real number. Calculate

$$\lim_{t \to \infty} \sqrt[n]{\sum_{k=0}^{n} \binom{n}{k}^t},$$

where $t$ is a real number.

**Exercise 2.6.18** Let $k$ be a non-negative integer. Show that any natural number $n$ can be written uniquely as

$$n = \binom{x_k}{k} + \binom{x_{k-1}}{k-1} + \cdots + \binom{x_1}{1},$$

where $1 \le x_1 < x_2 < \cdots < x_k$.

**Exercise 2.6.19** Let $B_n$ denote the $n$th Bell number. Show that $B_n < n!$ for each $n \ge 3$.

**Exercise 2.6.20** Determine the number of ways of writing a natural number $n$ as a sum of ones and twos.

# References

K. Plofker, *Mathematics in India* (Princeton University Press, Princeton, 2009)
G. Joseph, *The Crest of the Peacock, Non-European Roots of Mathematics* (Penguin Books, 1991)

# Chapter 3
# The Principle of Inclusion and Exclusion

## 3.1 The Main Theorem

The principle of inclusion and exclusion was used by the French mathematician Abraham de Moivre (1667–1754) in 1718 to calculate the number of derangements on $n$ elements. Since then, it has found innumerable applications in many branches of mathematics. It is not only an essential principle in combinatorics but also in number theory, especially with respect to sieve methods.

Let $A$ be a finite set and $A_1, \ldots, A_n \subseteq A$ be some subsets of $A$. We would like to know how many elements there are in the set

$$A \setminus \cup_{j=1}^{n} A_j.$$

One way to think about this situation in applications is to consider $A$ as some universal set of objects, and the subsets $A_1, \ldots, A_n$ be some *bad* subsets we are trying to avoid.

To state the principle of inclusion and exclusion, we will use the following notation. For each subset $J$ of $\{1, 2, \ldots, n\}$, let

$$A_J = \cap_{j \in J} A_j.$$

We define $A_\emptyset$ as $A$. The principle of inclusion-exclusion is contained in the following theorem.

**Theorem 3.1.1** *The number of elements not belonging to any $A_j$, $1 \leq j \leq n$ is given by*

$$\sum_{J \subseteq \{1,2,\ldots,n\}} (-1)^{|J|} |A_J|.$$

© Hindustan Book Agency 2022

S. M. Cioabă and M. R. Murty, *A First Course in Graph Theory and Combinatorics*,
Texts and Readings in Mathematics 55, https://doi.org/10.1007/978-981-19-0957-3_3

***Proof*** The sum is equal to

$$\sum_{J}(-1)^{|J|}\sum_{a\in A_J}1 = \sum_{a\in A}\sum_{J:a\in A_J}(-1)^{|J|}.$$

Let $S_a$ be the set of indices $j$ such that $a \in A_j$. The inner sum is a sum over all subsets of $S_a$. If $S_a = \emptyset$, this sum is 1. Otherwise, by the binomial theorem, it is equal to

$$\sum_{j=0}^{|S_a|}(-1)^j\binom{|S_a|}{j} = (1-1)^{|S_a|} = 0.$$

Hence, the sum is equal to the number of elements $a$ for which $S_a$ is empty. Since the number of elements not belonging to any $A_j$ is precisely the number of elements $a$ for which $S_a$ is empty, this completes the proof. ∎

This simple principle is one of the most powerful in all of mathematics and has profound consequences, which we will present in the next sections.

## 3.2   Derangements Revisited

It will be recalled that in Chap. 2, we derived a recurrence relation for the number of derangements of a set with $n$ elements. We used that recurrence relation to determine the exponential generating function of the number of derangements which lead to Theorem 2.4.2. We now give a simpler proof of that formula. Let $A$ be the set $S_n$ of all the permutations of $n$ elements. The number $d_n$ counts the number of permutations without any fixed points. For each $k$, $1 \le k \le n$, let $A_k$ be the subset of *bad* permutations that fix $k$. The number of derangements is the number of permutations not belonging to any of the $A_k$, $1 \le k \le n$. For each subset $J$ of $\{1, 2, .., n\}$, the number of elements of $A_J$ is clearly $(n - |J|)!$. As there are $\binom{n}{j}$ such subsets with $|J| = j$, by the principle of inclusion and exclusion, we obtain that

$$d_n = \sum_{j=0}^{n}(-1)^j\binom{n}{j}(n-j)!.$$

This is precisely the formula we established in Theorem 2.4.2. One advantage of using the principle of inclusion and exclusion is that the argument above can be generalized to count the number of permutations having exactly $r$ fixed points for any given integer $r$ between 1 and $n$ (see Exercise 3.6.7).

## 3.3 Counting Surjective Maps

Let us now count the number of surjective functions $f : [n] \to [k]$.

**Theorem 3.3.1** *The number of surjective functions $f : [n] \to [k]$ is*

$$\sum_{j=0}^{k}(-1)^{j} \binom{k}{j}(k - j)^{n}.$$

**Proof** Our *universal set A* is the collection of all functions $f : [n] \to [k]$ and, clearly, $|A| = k^{n}$. Our *bad sets* are the following. For each $1 \le j \le k$, let $A_{j}$ be the set of maps $f : [n] \to [k]$ such that $f(x) \ne j$ for any $x \in [n]$. It is not too hard to see that $|A_{j}| = (k - 1)^{n}$ and more generally, $|A_{J}| = (k - |J|)^{n}$ for any $J \subseteq [k]$. By the principle of inclusion and exclusion, the result is now immediate. ∎

**Corollary 3.3.2** *If $k$ and $n$ are non-negative integers, then*

$$\sum_{j=0}^{n}(-1)^{j} \binom{k}{j}(n - j)^{n} = \begin{cases} 0 & \text{if } k < n \\ n! & \text{if } k = n. \end{cases}$$

**Proof** If $n < k$, there are no surjective functions from $[n]$ to $[k]$. If $n = k$, there are exactly $n!$ surjective functions from $[n]$ to $[k]$. The result now follows from the previous theorem. ∎

## 3.4 Stirling Numbers of the First Kind

We now introduce **Stirling numbers of the first kind**, denoted by $s(n, k)$. Recall that any permutation has a unique decomposition (up to rearrangement) as a product of disjoint cycles. We define $s(n, k)$ by the rule that $(-1)^{n-k}s(n, k)$ is the number of permutations of $S_{n}$ which can be written as a product of $k$-disjoint cycles. Clearly, $s(n, n) = 1$ since the only permutation that can have $n$ disjoint cycles in its cycle decomposition is the identity permutation. It is also clear that

$$\sum_{k=1}^{n}(-1)^{n-k}s(n, k) = \sum_{k=1}^{n}|s(n, k)| = n!$$

We now establish a recurrence for $s(n, k)$.

**Theorem 3.4.1** *For any natural numbers $n \ge k$,*

$$s(n + 1, k) = -ns(n, k) + s(n, k - 1).$$

**Proof** Of the permutations of $S_{n+1}$ with $k$ disjoint cycles, we consider those in which $(n + 1)$ appears as a one cycle and those in which it does not. The number in the first group is clearly $(-1)^{n-(k-1)}s(n, k - 1)$. For the number in the second group, we may view the elements as permutations of $S_n$ with $k$ disjoint cycles into which we have *interpolated* $(n + 1)$. For a cycle of $S_n$ of length $j$, there are $j$ places into which we can insert $(n + 1)$, giving $j$ new permutations. Now if $\sigma$ is a permutation of $S_n$ with $k$-cycles of lengths $j_1, \ldots, j_k$, we can interpolate $(n + 1)$ into this in $j_1 + \cdots + j_k = n$ ways. Thus, the number of elements in the second group is $n(-1)^{n-k}s(n, k)$.

Thus,

$$(-1)^{n+1-k}s(n + 1, k) = (-1)^{n-(k-1)}s(n, k - 1) + n(-1)^{n-k}s(n, k).$$

This simplifies to give the stated recurrence.  ∎

For a real number $t$ and a natural number $n$, we denote $(t)_n = t(t - 1) \ldots (t - n + 1)$. Using the previous result, we can prove the following result.

**Theorem 3.4.2** *For any real number $t$ and any natural number $n$,*

$$(t)_n = \sum_{k=1}^{n} s(n, k)t^k.$$

**Proof** Again, we use induction. For $n = 1$, the result is clear as $s(1, 1) = 1$. Assume that the result is established for some $n \geq 1$. We show that the equation above is true for $n + 1$. Note that

$$(t)_{n+1} = (t)_n(t - n) = \left( \sum_{k=1}^{n} s(n, k)t^k \right) \cdot (t - n).$$

The coefficient of $t^k$ on the right side is $s(n, k - 1) - ns(n, k)$ which is $s(n + 1, k)$ by the previous theorem. This completes the proof.  ∎

## 3.5   Stirling Numbers of the Second Kind

We study now the **Stirling numbers of the first kind**, denoted by $S(n, k)$. For natural numbers $n \geq k \geq 1$, $S(n, k)$ equals the number of partitions of an $[n]$ into $k$ parts or blocks. We first relate these numbers to the discussion of the surjective functions. Observe that if we have a surjective map $f : [n] \rightarrow [k]$, the pre-images $f^{-1}(y) = \{x \in [n] : f(x) = y\}$, for $1 \leq y \leq k$, form a partition of the $[n]$ into $k$ blocs. Conversely, given a partition of $[n]$ into $k$ parts, there are $k!$ ways of defining a surjective function $f : [n] \rightarrow [k]$ since we can view each block as the fiber of the image of such a map and there are $k!$ ways of pairing up the $k$ blocks with the elements of $[k]$. Putting this together with Theorem 3.3.1 gives the next result.

**Theorem 3.5.1** *For any natural numbers $n \geq k$,*

$$k!S(n,k) = \sum_{j=0}^{k}(-1)^j \binom{k}{j}(k-j)^n.$$

This formula allows us to deduce the generating function for the Bell numbers $B_n$ that we derived in the previous chapter. Indeed, we have that

$$B_n = \sum_{k=0}^{n} S(n,k).$$

On the other hand, notice that

$$\sum_{n\geq 0} \frac{S(n,k)t^n}{n!} = \frac{1}{k!}\sum_{n\geq 0}\frac{t^n}{n!}\sum_{j=0}^{k}(-1)^j\binom{k}{j}(k-j)^n.$$

Upon interchanging the summation, we get that this equals

$$\frac{1}{k!}\sum_{j=0}^{k}(-1)^j\binom{k}{j}e^{(k-j)t}.$$

Using the binomial theorem, we can simplify the right-hand side and deduce the following result.

**Theorem 3.5.2** *For any natural number $k$,*

$$\sum_{n\geq 0}\frac{S(n,k)t^n}{n!} = \frac{1}{k!}(e^t-1)^k.$$

Combining this fact with the formula relating $B_n$ with the Stirling numbers of the second kind easily gives us again the generating function

$$\sum_{n\geq 0}\frac{B_n t^n}{n!} = e^{e^t-1}.$$

Even though we have an explicit formula for the $S(n,k)$'s, it will be useful to derive the following recurrence relation.

**Theorem 3.5.3** *For any natural numbers $n \geq k$,*

$$S(n,k) = S(n-1,k-1) + kS(n-1,k).$$

***Proof*** In partitioning the $n$-set $\{1, 2, \ldots, n\}$ into $k$ blocks, we have two possibilities. Either $n$ is in a singleton block or it is not. In the first case, the number of such decompositions clearly corresponds to $S(n-1, k-1)$. In the second case, we take the decomposition of a $(n-1)$-set into $k$-blocks, and we now have $k$ choices into which we may place $n$. This gives the recursion. ∎

We may use this recursion to give another interpretation of the numbers $S(n, k)$. To this end, we recall the notation $(t)_n = t(t-1)(t-2)\ldots(t-n+1)$.

**Theorem 3.5.4** *For any real number $t$ and any natural number $n$,*

$$t^n = \sum_{k=1}^{n} S(n, k)(t)_k.$$

***Proof*** The proof is again by induction on $n$. For $n = 1$, the result is clear as $S(1, 1) = 1$. Let $n \geq 1$ and suppose that we have proved the formula for $n$. We will now prove it for $n + 1$. We have that

$$t^{n+1} = t^n \cdot t = \sum_{k=1}^{n} S(n, k)(t)_k((t-k) + k)$$

by the induction hypothesis. Since $(t)_k(t-k) = (t)_{k+1}$, we deduce that

$$t^{n+1} = \sum_{k=1}^{n} S(n, k)(t)_{k+1} + \sum_{k=1}^{n} k S(n, k)(t)_k.$$

By changing variables on the first sum, and noting that $S(n, n+1) = 0$, we may write the right-hand side as

$$\sum_{k=1}^{n+1} \{S(n, k-1) + k S(n, k)\}(t)_k = \sum_{k=1}^{n+1} S(n+1, k)(t)_k$$

by the recursion of Theorem 3.5.3. This completes the proof. ∎

**Corollary 3.5.5** *Let $n$ be a natural number. If $A$ and $B$ are the $n \times n$ matrices whose $(i, j)$th entries are given by $s(i, j)$ and $S(i, j)$, respectively, then $B = A^{-1}$.*

***Proof*** Let $\mathbb{R}[t]$ denote the collection of polynomials with real coefficients in variable/indeterminate $t$. Denote by $V$ the vector space of polynomials in $\mathbb{R}[t]$ of degree $n$ or less, with constant term zero. Then $A$ and $B$ are the transition matrices between the following two bases of $V$:

(1) $t, \ldots, t^n$,
(2) $(t)_1, \ldots, (t)_n$.

The result now follows from linear algebra. ∎

**Corollary 3.5.6** *Let $(f_n)_{n \geq 1}$ and $(g_n)_{n \geq 1}$ be two sequences of real numbers. For a natural number n, the following statements are equivalent.*

(1)  $g_n = \sum_{k=1}^{n} S(n, k) f_k$;
(2)  $f_n = \sum_{k=1}^{n} s(n, k) g_k$.

**Proof**  This is immediate from matrix inversion.                                    ∎

If we define $f_0$ and $g_0$ so that $f_0 = g_0$, and

$$F(t) = \sum_{n \geq 0} \frac{f_n t^n}{n!}$$

$$G(t) = \sum_{n \geq 0} \frac{g_n t^n}{n!},$$

where $g_n$ and $f_n$ are related as in Corollary 3.5.6, then we can determine the relationship between these two generating functions as follows:

$$G(t) = f_0 + \sum_{n \geq 1} \sum_{k=1}^{n} S(n, k) f_k \frac{t^n}{n!}.$$

Interchanging summations, we deduce that

$$G(t) = f_0 + \sum_{k \geq 1} f_k \frac{(e^t - 1)^k}{k!} = F(e^t - 1).$$

This proves our next result.

**Corollary 3.5.7**  *If $f_n$ and $g_n$ are related as in Corollary 3.5.6, then*

$$G(t) = F(e^t - 1).$$

This allows us to deduce the generating function for Stirling numbers of the first kind. Let $g_k = 1$ and $g_n = 0$ for $n \neq k$. Then $f_n = s(n, k)$. By Corollary 3.5.7, we get that $\frac{t^k}{k!} = F(e^t - 1)$. Putting $x = e^t - 1$ gives that

$$\sum_{n \geq 0} \frac{s(n, k) x^k}{k!} = \frac{\log^k(1 + x)}{k!}.$$

## 3.6  Exercises

**Exercise 3.6.1**  There are 13 students taking math, 17 taking physics, and 18 taking chemistry. Also, there are 5 students taking both math and physics, 6 students taking physics and chemistry, and 4 taking chemistry and math. Only 2 students out of the total of 50 students are taking math, physics, and chemistry. How many students are not taking any courses at all?

**Exercise 3.6.2**  The **greatest common divisor** $\gcd(a, b)$ of two natural numbers $a$ and $b$ is the largest natural number that divides both $a$ and $b$. If $n$ is a natural number, denote by $\phi(n)$ the number of integers $k$ with $1 \leq k \leq n$ and $\gcd(n, k) = 1$. This is called **the Euler's totient function**.

(1)  Show that if $p$ is a prime and $\alpha$ is a natural number, then $\phi(p^\alpha) = p^\alpha - p^{\alpha-1}$.
(2)  If $\gcd(m, n) = 1$, then $\phi(mn) = \phi(m)\phi(n)$.

**Exercise 3.6.3**  Let $n \geq 2$ be a natural number whose decomposition as a product of distinct primes $n = p_1^{\alpha_1} \ldots p_k^{\alpha_k}$. Prove that

$$\phi(n) = n \prod_{i=1}^{k} \left( 1 - \frac{1}{p_i} \right).$$

**Exercise 3.6.4**  How many integers less than $n$ are not divisible by any of 2, 3, and 5?

**Exercise 3.6.5**  How many 7 digit phone numbers contain at least 3 odd digits?

**Exercise 3.6.6**  If $A_1, A_2, \ldots, A_n$ are finite sets, show that

$$\sum_{i=1}^{n} |A_i| - \sum_{i \neq j} |A_i \cap A_j| \leq |\cup_{i=1}^n A_i| \leq \sum_{i=1}^{n} |A_i|.$$

When does equality happen ?

**Exercise 3.6.7**  If $n$ and $r$ are non-negative integers with $0 \leq r \leq n$, denote by $f(n, r)$ the number of permutations of $S_n$ with exactly $r$ fixed points. Show that

$$\lim_{n \to \infty} \frac{f(n, r)}{n!} = \frac{1}{er!}.$$

**Exercise 3.6.8**  Let $(a_n)_{n \geq 0}$ be a sequence of real numbers with generating function $f(t) = \sum_{n \geq 0} a_n t^n$. Determine the generating functions of the sequences $(a_{n+1})_{n \geq 0}$ and $(na_n)_{n \geq 0}$ as a function of $f(t)$. Find the generating function for the sequence $(b_n)_{n \geq 0}$ defined by $b_0 = 1$ and $b_{n+1} = 2b_n + n$ for $n \geq 0$.

**Exercise 3.6.9**  Let $s(n, k)$ denote the Stirling numbers of the first kind. If $n$ is a natural number, show that

$$x(x + 1) \ldots (x + n - 1) = \sum_{k=0}^{n} |s(n, k)| x^k.$$

**Exercise 3.6.10** Using the previous identity, prove that the number of permutations with an even number of cycles (in their decomposition as a product of disjoint cycles) is equal to the number of permutations with an odd number of cycles.

**Exercise 3.6.11** Let $S(n, k)$ denote the Stirling numbers of second kind. If $n \geq k \geq 2$, show that

$$S(n + 1, k) = \sum_{j=1}^{n} \binom{n}{j} S(j, k - 1).$$

**Exercise 3.6.12** Prove that

$$\sum_{i=0}^{n} (-1)^i \binom{n}{i} \binom{m + n - i}{k - i} = \begin{cases} \binom{m}{k} & \text{if } m \geq k \\ 0 & \text{if } m < k. \end{cases}$$

**Exercise 3.6.13** Show that $|s(n, 1)| = (n - 1)!$. Give two proofs.

**Exercise 3.6.14** Prove that $S(n, 1) = S(n, n) = 1$, $S(n, 2) = 2^{n-1} - 1$, and $S(n, n - 1) = \binom{n}{2}$.

**Exercise 3.6.15** Show that for $1 \leq k \leq n$, $S(n, k) \geq \binom{n}{k-1}$.

**Exercise 3.6.16** For two natural numbers $k$ and $n$ define $p_k(n)$ as the number of integer solutions $(x_1, \ldots, x_k)$, $x_1 \geq \cdots \geq x_k \geq 1$ of

$$n = x_1 + \cdots + x_k.$$

For example, $7 = 5 + 1 + 1 = 4 + 2 + 1 = 3 + 3 + 1 = 3 + 2 + 2$ so $p_4(7) = 3$. For any given $k$, show that

$$\lim_{n \to \infty} \frac{p_k(n)}{\frac{n^{k-1}}{k!(k-1)!}} = 1.$$

**Exercise 3.6.17** The Bernoulli numbers $b_n$ are defined by the recurrence relation

$$\sum_{k=0}^{n} \binom{n + 1}{k} b_k = 0$$

for $n \geq 1$ and $b_0 = 1$. Prove that

$$g(t) := \sum_{n \geq 0} \frac{b_n t^n}{n!} = \frac{t}{e^t - 1}.$$

These numbers were studied by Jakob Bernoulli (1654–1705) in 1713.

**Exercise 3.6.18** Show that $g(t) + \frac{t}{2}$ is an even function of $t$, where $g(t)$ is defined in the previous exercise.

**Exercise 3.6.19** Show that $b_n = 0$ for each odd number $n \geq 3$.

**Exercise 3.6.20** Let $(f_n)_{n \geq 0}$ and $(g_n)_{n \geq 0}$ be sequences, with exponential generating functions $f(t) = \sum_{n \geq 0} \frac{f_n t^n}{n!}$ and $g(t) = \sum_{n \geq 0} \frac{g_n t^n}{n!}$. Show that the following two statements are equivalent:

(1)  For any natural number $n$, $g_n = \sum_{k=0}^{n} \binom{n}{k} f_k$
(2)  $g(t) = e^t f(t)$.

# Chapter 4
# Graphs and Matrices

## 4.1 Adjacency Matrix

Given a graph, one can associate various matrices to encode its information. **The adjacency matrix** $A$ of a graph $X = (V, E)$ is the matrix whose rows and columns are indexed by the vertices of $X$, where $A(x, y)$ equals the number of edges between $x$ and $y$. When necessary to indicate the dependence on $X$, we denote $A$ by $A(X)$. A number $\lambda$ is called **an eigenvalue** of $A$ if there exists a non-zero vector $u \in \mathbb{R}^n$ such that $Au = \lambda u$. This means that $\lambda$ is a root of **the characteristic polynomial** $\det(t I_n - A)$. If $n$ denotes the number of vertices of $X$, then $A$ is a $n \times n$ real and symmetric matrix and therefore by linear algebra, it has $n$ real eigenvalues which we denote as follows:

$$\lambda_1 \geq \cdots \geq \lambda_n.$$

**The spectrum of** $X$ consists of its eigenvalues, including multiplicities.

**Example 4.1.1** In Fig. 4.1, we describe two graphs.

The adjacency matrices of these graphs are below.

$$A(K_4) = \begin{bmatrix} 0 & 1 & 1 & 1 \\ 1 & 0 & 1 & 1 \\ 1 & 1 & 0 & 1 \\ 1 & 1 & 1 & 0 \end{bmatrix}, \quad A(K_{1,3}) = \begin{bmatrix} 0 & 0 & 0 & 1 \\ 0 & 0 & 0 & 1 \\ 0 & 0 & 0 & 1 \\ 1 & 1 & 1 & 0 \end{bmatrix}.$$

The eigenvalues of these matrices are the following:

$$A(K_4) : 3, -1, -1, -1$$
$$A(K_{1,3}) : \sqrt{3}, 0, 0, -\sqrt{3}.$$

The original version of this chapter was revised: Figure 4.1 in Chapter 4 has been updated. The correction to this chapter is available at https://doi.org/10.1007/978-981-19-0957-3_14

© Hindustan Book Agency 2022, corrected publication 2022
S. M. Cioabă and M. R. Murty, *A First Course in Graph Theory and Combinatorics*,
Texts and Readings in Mathematics 55, https://doi.org/10.1007/978-981-19-0957-3_4

Fig. 4.1 The graphs $K_4$ and $K_{1,3}$

We will see later in the chapter how to calculate these eigenvalues for any complete graph and any complete bipartite graph. Our first result describes a combinatorial meaning for the entries of the powers of the adjacency matrix.

**Theorem 4.1.2** *Let $X = (V, E)$ be a graph without loops, with adjacency matrix $A$. For any $x, y \in V$ and any natural number $\ell$, $A^{\ell}(x, y)$ equals the number of $(x, y)$-walks of length $\ell$.*

***Proof*** Denote by $w_{\ell}(x, y)$ the number of $(x, y)$-walks of length $\ell$. We will show that $A^{\ell}(x, y) = w_{\ell}(x, y)$ for any $x, y \in V$ and any natural number $\ell$ using induction on $\ell$. The base case $\ell = 1$ is clear from the definition of $A$. Let $\ell \geq 2$ now and assume that $A^{\ell-1}(x, y) = w_{\ell-1}(x, y)$ for any $x, y \in V$. Let $x$ and $y$ be two vertices of $X$. A $(x, y)$-walk of length $\ell$ is composed of a $(x, z)$-walk of length $\ell - 1$ and an edge $zy$. Hence,

$$w_{\ell}(x, y) = \sum_{z:z \sim y} w_{\ell-1}(x, z).$$

At the same time,

$$A^{\ell}(x, y) = \sum_{z \in V} A^{\ell-1}(x, z) A(z, y) = \sum_{z:z \sim y} A^{\ell-1}(x, z).$$

From the induction hypothesis, we know that $A^{\ell-1}(x, z) = w_{\ell-1}(x, z)$ for each $z \sim y$. Using the above two equations, we get that $A^{\ell}(x, y) = w_{\ell}(x, y)$ which finishes our proof. ∎

A consequence of the previous result is the following corollary, which shows a close connection between the eigenvalues of a graph and the number of its closed walks. **The trace** tr $(M)$ of a square matrix $M$ is the sum of its diagonal entries.

**Corollary 4.1.3** *Let $X = (V, E)$ be a graph with adjacency matrix $A$ whose eigenvalues are $\lambda_1 \geq \cdots \geq \lambda_n$. For any natural number $\ell$, the number of closed walks of length $\ell$ in $X$ equals $\sum_{j=1}^{n} \lambda_j^{\ell}$.*

***Proof*** The eigenvalues of $A^{\ell}$ are $\lambda_1^{\ell}, \ldots, \lambda_n^{\ell}$ and

$$\text{tr}\,(A^{\ell}) = \sum_{j=1}^{n} \lambda_j^{\ell}.$$

On the other hand, the trace of $A^{\ell}$ is the sum of the diagonal entries of $A^{\ell}$. By Theorem 4.1.2, this equals the number of closed walks of length $\ell$ in $X$. ∎

**Fig. 4.2** Two isomorphic graphs

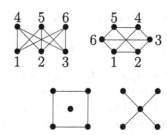

**Fig. 4.3** Two non-isomorphic and cospectral graphs

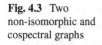

## 4.2  Graph Isomorphisms

An **isomorphism** between two graphs $X = (V, E)$ and $Y = (W, F)$ is a bijection $f : V \to W$ between the vertex set of $X$ and the vertex set of $Y$ that maps edges to edges and non-edges to non-edges:

$$xy \in E \Leftrightarrow f(x)f(y) \in F.$$

We will only consider isomorphisms of simple graphs. Informally speaking, when $V = W$, a graph isomorphism is a relabeling of the vertices of the graph. More formally, this means that applying a permutation to both the rows and columns of the adjacency matrix has the effect of reordering the vertices of $X$.

**Theorem 4.2.1** *Let $X$ and $Y$ be two simple graphs on the same vertex set $V$. The graphs $X$ and $Y$ are isomorphic if and only if there is a $V \times V$ permutation matrix $P$ such that*

$$A(Y) = PA(X)P^{-1}.$$

The reader is invited to show that the graphs in Fig. 4.2 are isomorphic and determine the matrix $P$ in this case.

Thus, for two graphs to be isomorphic, it is necessary that their adjacency matrices are cospectral meaning that they have the same eigenvalues. However, this is not a sufficient condition for isomorphism, as the following remark shows.

Consider the graph obtained from $C_4$ by adding an isolated vertex. This graph has the same eigenvalues as $K_{1,4}$, but it is not isomorphic to $K_{1,4}$ (see Exercise 4.6.15) (Fig. 4.3).

**Example 4.2.2** As a warm-up, let us compute the eigenvalues and the characteristic polynomial of the $n \times n$ matrix $J_n$ whose entries are all one. This matrix has rank 1 and therefore 0 is an eigenvalue with multiplicity $n - 1$. Also, the all one vector of dimension $n$ is an eigenvector of $J_n$ with eigenvalue $n$. The characteristic polynomial of $J_n$ is $t^{n-1}(t - n)$.

**Example 4.2.3** We now determine the eigenvalues of the complete graph $K_n$. The adjacency matrix of $K_n$ is $J_n - I_n$ with $J_n$ defined above in Example 4.2.2. Note that if $A$ has eigenvalue $\mu$, then $\mu + c$ is an eigenvalue of $A + cI_n$. Thus, the eigenvalues of $J_n - I_n$ are $n - 1$ and $-1$ with multiplicity 1 and $n - 1$, respectively. The characteristic polynomial of the adjacency matrix of $K_n$ is $(t - (n - 1))(t + 1)^{n-1}$.

**Example 4.2.4** We determine the eigenvalues of the adjacency matrix of the bipartite graph $K_{r,s}$. For natural numbers $r$ and $s$, denote by $O_r$ the $r \times r$ matrix with all entries zero and by $J_{r,s}$ the $r \times s$ matrix with all entries one. Assume that the partite sets of this graph are labeled $1, \ldots, r$ and $r + 1, \ldots, r + s$, respectively. The adjacency matrix of $K_{r,s}$ has the following block-matrix form:

$$A(K_{r,s}) = \begin{bmatrix} O_r & J_{r,s} \\ J_{s,r} & O_s \end{bmatrix}.$$

This matrix has rank two and therefore, 0 is an eigenvalue with multiplicity $r + s - 2$. If we denote the remaining two eigenvalues by $\theta \geq \tau$, then $\theta + \tau = 0$ because tr $(A_{r,s}) = 0$. Corollary 4.1.3 tells us that $\theta^2 + \tau^2 = \text{tr}(A^2)$ equals the number of closed walks of length two in $K_{r,s}$ which is twice the number of edges (see Exercise 4.6.3) of $K_{r,s}$. Putting these things together, we get that $\theta = \sqrt{rs}$ and $\tau = -\sqrt{rs}$. The characteristic polynomial of $K_{r,s}$ is $t^{r+s-2}(t^2 - rs)$.

## 4.3  Bipartite Graphs and Eigenvalues

The fact that the spectrum of the $K_{r,s}$ is symmetric with respect to zero is a consequence of the following general result regarding the spectra of bipartite graphs.

**Theorem 4.3.1** *If $X$ is bipartite, and $\lambda > 0$ is an eigenvalue of the adjacency matrix of $X$ with multiplicity $m$, then $-\lambda$ is also an eigenvalue of multiplicity $m$.*

**Proof** Assume that $X$ is a bipartite graph whose partite sets have sizes $r$ and $s$, respectively. We label our vertices similar to Example 4.2.4 such that the adjacency matrix $A$ of $X$ has the following form:

$$A = \begin{bmatrix} O_r & B \\ B^T & O_s \end{bmatrix},$$

where $B$ is a $r \times s$ matrix and $B^T$ is its transpose. If $\lambda > 0$ is an eigenvalue of $A$ with an eigenvector

$$w = \begin{bmatrix} u \\ v \end{bmatrix}$$

partitioned according to the partite sets (so $u$ is a $r$-dimensional real vector and $v$ is a $s$-dimensional real vector, not both of them being zero), then

$$\lambda \begin{bmatrix} u \\ v \end{bmatrix} = \lambda w = Aw = \begin{bmatrix} O_r & B \\ B^T & O_s \end{bmatrix} \begin{bmatrix} u \\ v \end{bmatrix}.$$

This implies that $\lambda u = Bv$ and $\lambda v = B^T u$. If we define now the vector $w' = \begin{bmatrix} u \\ -v \end{bmatrix}$, we get that

$$Aw' = \begin{bmatrix} O_r & B \\ B^T & O_s \end{bmatrix} \begin{bmatrix} u \\ -v \end{bmatrix} = \begin{bmatrix} -Bv \\ B^T u \end{bmatrix} = \begin{bmatrix} -\lambda u \\ \lambda v \end{bmatrix} = (-\lambda)w'.$$

The vector $w'$ is not the zero vector because $w$ was not the zero vector. Thus, $w'$ is an eigenvector of $A$ with eigenvalue $-\lambda$. A not too difficult argument shows that a collection of $m$ linearly independent eigenvectors corresponding to $\lambda$ will give us a set of $m$ linearly independent eigenvectors corresponding to $-\lambda$. This implies that the multiplicity of $-\lambda$ is at least the multiplicity of $\lambda$. Switching $\lambda$ with $-\lambda$ and repeating our previous argument shows us that the multiplicities of these eigenvalues are equal. ∎

We give now a spectral characterization of bipartite graphs.

**Theorem 4.3.2** *Let $X$ be a graph on $n$ vertices with adjacency matrix $A$ and eigenvalues $\lambda_1 \geq \cdots \geq \lambda_n$. The following statements are equivalent.*

*(1) The graph $X$ is bipartite.*
*(2) The spectrum of $X$ is symmetric with respect to zero.*
*(3) If $n$ is even, the characteristic polynomial $\det(t I_n - A)$ of $A$ is a polynomial in $t^2$ and if $n$ is odd, $t^{-1} \det(t I_n - A)$ is a polynomial in $t^2$.*
*(4) For any odd natural number $\ell$, $\sum_{j=1}^{n} \lambda_j^\ell = 0$.*

**Proof** The fact that (1) implies (2) was done in the previous theorem. To see (2) implies (3), note that when $n$ is even, for any positive eigenvalue $\lambda$ of multiplicity $m$, we will have the eigenvalue $-\lambda$ of same multiplicity $m$. The factors $(t - \lambda)^m (t + \lambda)^m$ can combine into $(t^2 - \lambda^2)^m$, a polynomial in $\lambda^2$. The eigenvalue 0 will have an even (possibly zero) multiplicity and, therefore, $\det(t I_n - A)$ will be a polynomial in $t^2$. When $n$ is odd, the argument is similar, with the difference that the multiplicity of 0 in this case will be odd. We leave the implications (3) implies (2) and (2) implies (4) as homework for the interested reader. To see that (4) implies (1), note that condition (4) means that $X$ has no closed odd walks and therefore, no odd cycles. Hence, $X$ is bipartite. The theorem is proved. ∎

## 4.4 Diameters and Eigenvalues

**The diameter** of a connected graph $X = (V, E)$ is defined as

$$\mathrm{diam}(X) = \max_{x,y \in V} d(x, y).$$

**The minimal polynomial** of a real and symmetric $n \times n$ matrix $M$ is the monic polynomial $m(t) \in \mathbb{R}[t]$ with real coefficients of the smallest degree such that $m(M) = O_n$. It equals $(t - \theta_1) \ldots (t - \theta_r)$, where $\theta_1, \ldots, \theta_r$ are the distinct eigenvalues of $M$. If $A$ is the adjacency matrix of $X$, the set of polynomials $\mathcal{A}(X)$ in $A$ with real coefficients is a vector space over the set of real numbers. It is called **the adjacency algebra of** $X$ and its dimension equals $r$.

**Theorem 4.4.1** *If $X$ is a connected graph whose adjacency matrix has $r$ distinct eigenvalues, then* $\mathrm{diam}(X) \leq r - 1$.

***Proof*** Let $d = \mathrm{diam}(X)$. We show that $I_n, A, \ldots, A^d$ are linearly independent matrices. Because $\dim \mathcal{A}(X) = r$, this implies that $d + 1 \leq r$ which proves the theorem.

To show that $I_n, A, \ldots, A^d$ are linearly independent, we will prove that for any $1 \leq k \leq d$, $A^k$ cannot be written as a linear combination of $I_n, \ldots, A^{k-1}$. Consider two vertices $x, y \in V$ such that $d(x, y) = k$. Therefore, there exists a $(x, y)$-path of length $k$ and there are no $(x, y)$-walks of length $\ell$ for any $\ell < k$. Theorem 4.1.2 implies that $A^k(x, y) > 0$ and $A^\ell(x, y) = 0$ for any $\ell < k$. Therefore, $A^k$ cannot be written as a linear combination of $I_n, \ldots, A^{k-1}$. This finishes our proof.    ∎

The examples in the previous section show that this result is sharp. For instance, in the case of the complete graph $K_n$, the diameter is one and the number of distinct eigenvalues is two. In the case of the bipartite graph $K_{r,s}$, we have a diameter of two and the number of distinct eigenvalues is three.

## 4.5  Incidence Matrices and the Laplacian Matrix

In this section, we discuss other matrices associated with a graph and their basic properties. In a simple graph $X = (V, E)$, we choose an **orientation** of the edges by assigning a direction to each edge. This way, for each edge $e$, one of its endpoints becomes **the tail** and the other endpoint is **the head** of the oriented edge. Of course, their roles switch if we reverse the orientation. Since each edge has two possible orientations, there are $2^{|E|}$ possible orientations of $X$. Given an orientation of $X$, **the oriented incidence matrix** or **the directed incidence matrix** $N$ is defined as follows. Its rows are labeled by the vertex set $V$ and the columns by the edge set $E$. If $x \in V, e \in E$, then

$$N(x, e) = \begin{cases} +1, & \text{if } x \text{ is the head of } e \\ -1, & \text{if } x \text{ is the tail of } e \\ 0, & \text{otherwise.} \end{cases}$$

**Example 4.5.1** Figure 4.4 describes the orientation of a graph on four vertices and five edges. The tip of each arrow is the head and the other endpoint is the tail.

**Fig. 4.4** An orientation of a
graph on four vertices

In the case when the vertices are listed in the order $1, 2, 3, 4$ and the edges are labelled in the order $12, 13, 14, 23, 34$, the oriented incidence matrix is below:

$$N = \begin{bmatrix} -1 & 1 & -1 & 0 & 0 \\ 1 & 0 & 0 & -1 & 0 \\ 0 & -1 & 0 & 1 & -1 \\ 0 & 0 & 1 & 0 & 1 \end{bmatrix}.$$

For a graph $X = (V, E)$, denote by $D$ be the $V \times V$ diagonal matrix whose $(x, x)$th entry equals the degree of the vertex $x$. We call $D$ **the diagonal degree matrix** of $X$.

Note that

$$NN^T = \begin{bmatrix} 3 & -1 & -1 & -1 \\ -1 & 2 & -1 & 0 \\ -1 & -1 & 3 & -1 \\ -1 & 0 & -1 & 2 \end{bmatrix} = D - A,$$

where

$$D = \begin{bmatrix} 3 & 0 & 0 & 0 \\ 0 & 2 & 0 & 0 \\ 0 & 0 & 3 & 0 \\ 0 & 0 & 0 & 4 \end{bmatrix},$$

is the diagonal degree matrix of $X$ and

$$A = \begin{bmatrix} 0 & 1 & 1 & 1 \\ 1 & 0 & 1 & 0 \\ 1 & 1 & 0 & 1 \\ 1 & 0 & 1 & 0 \end{bmatrix}$$

is the adjacency matrix of $X$.

**The Laplacian matrix** $L$ of the graph $X$ is defined as $L = D - A$, where $D$ is the diagonal degree matrix of $X$ and $A$ is the adjacency matrix of $X$. This is another important matrix associated to a graph, and we summarize its basic properties below.

**Theorem 4.5.2** *Let $X = (V, E)$ be a graph and $N$ an oriented incidence matrix of it.*

*(1) The Laplacian matrix $L$ equals $NN^T$.*

(2) *The Laplacian matrix L is real and symmetric, and its eigenvalues are non-negative.*
(3) *The smallest eigenvalue of the Laplacian matrix L is 0, and its multiplicity equals the numbers of components of X.*

**Proof** For part (1), let $x, y \in V$. If $x = y$, the $(x, x)$th entry in $L$ is $d(x)$. At the same time,

$$(NN^T)(x, x) = \sum_{e \in E} N(x, e)N^T(e, x) = \sum_{e \in E} N^2(x, e) = d(x).$$

If $x \neq y$ and $x \sim y$, then $L(x, y) = -1$. If $f$ is the edge connecting $x$ to $y$, then regardless of its orientation, $N(x, f)N(y, f) = -1$. For any other edge $e \neq f$, $N(x, e)N(y, e) = 0$ and, therefore,

$$(NN^T)(x, y) = \sum_{e \in E} N(x, e)N(y, e) = -1.$$

If $x \neq y$ are not adjacent, then $L(x, y) = 0$. In this case, for any edge $e \in E$, $N(x, e)N(y, e) = 0$ and therefore,

$$(NN^T)(x, y) = \sum_{e \in E} N(x, e)N(y, e) = 0.$$

Hence, $L = NN^T$.

For part (2), since $D$ and $A$ are real and symmetric and $L = D - A$, it is clear that $L$ is real and symmetric. Let $\mu$ be an eigenvalue of $L$ with a corresponding eigenvector $u$. Then $\mu u = LU$ means that $\mu u = NN^T u$. Multiplying to the left by $u^T$, we get that $\mu u^T u = u^T NN^T u = (N^T u)^T (N^T u)$. Since $u^T u = \sum_{x \in V} u_x^2 > 0$ (as $u$ is a non-zero vector), we get that

$$\mu = \frac{(N^T u)^T (N^T u)}{u^T u} = \frac{\sum_{xy \in E} (u_x - u_y)^2}{\sum_{x \in V} u_x^2} \geq 0.$$

For part (3), consider the all vector $\mathbf{1}$ of dimension $n$. It is not too hard to see that $L\mathbf{1}$ is the zero vector of dimension $n$ since each row of $L$ sums up to 0. Hence, 0 is an eigenvalue of $L$ and $\mathbf{1}$ is an eigenvector corresponding to it.

We show that if $X$ is connected, the multiplicity of 0 is one. Let $u$ be an eigenvector corresponding to 0. From the previous arguments, it follows that

$$0 = \frac{\sum_{xy \in E} (u_x - u_y)^2}{\sum_{x \in V} u_x^2}.$$

This means that $u_x = u_y$ for any edge $xy$ of $X$. We leave it as an exercise for the motivated reader to deduce how this implies that $u_x = u_y$ for any $x, y \in V$. This means that $u$ is a multiple of the all vector $\mathbf{1}$ and therefore, the dimension of the eigenspace of 0 is one, finishing our proof.

Part (3) follows by noticing that the Laplacian of $X$ can be written as a block diagonal matrix, with its diagonal blocks being the Laplacians of its components. The result then follows from our argument above.                                    ∎

Note that while the spectrum of the adjacency matrix of a graph can tell us if the graph is bipartite, it cannot tell us if the graph is connected, as the examples in Fig. 4.3 show. However, the eigenvalues of the Laplacian matrix indicate when the graph is connected, as the above result shows.

**The unoriented incidence matrix** of the graph $X = (V, E)$ can be defined as follows. As before, the rows of $M$ are indexed by the vertex set $V$ and the columns are labeled by the edge set $E$. For $x \in V, e \in E$,

$$M(x, e) = \begin{cases} 1, & \text{if } x \text{ is an endpoint of } e \\ 0, & \text{otherwise.} \end{cases}$$

The relationship between this matrix and the adjacency matrix is given by (the easily verified)

$$MM^T = D + A,$$

where $D$ is the diagonal degree matrix and $A$ is the adjacency matrix of $X$. The matrix $Q := D + A$ is called **the signless Laplacian** of $X$.

## 4.6  Exercises

**Exercise 4.6.1** Show that a graph $X$ with $n$ vertices is connected if and only if $(A + I_n)^{n-1}$ has no zero entries, where $A$ is the adjacency matrix of $X$.

**Exercise 4.6.2** Let $X$ be a $k$-regular graph. Show that $\lambda$ is an eigenvalue of its adjacency matrix if and only if $k - \lambda$ is an eigenvalue of its Laplacian matrix.

**Exercise 4.6.3** Let $X$ be a $k$-regular graph.

(1) Show that $k$ is the largest eigenvalue of the adjacency matrix $A$ of $X$.
(2) Prove that $X$ is connected if and only if the multiplicity of $k$ is one.

**Exercise 4.6.4** Let $X$ be a $k$-regular graph.

(1) If $\lambda$ is an eigenvalue of the adjacency matrix of $X$, show that $\lambda \geq -k$.
(2) If $X$ is connected and $k$-regular, then prove that $X$ is bipartite if and only if $-k$ is an eigenvalue of its adjacency matrix.

**Exercise 4.6.5** Let $X$ be a simple graph with $e$ edges and $t_3$ triangles. If $A$ is the adjacency matrix of $X$, show that

$$\text{tr}(A) = 0, \ \text{tr}(A^2) = 2e, \ \text{tr}(A^3) = 6t_3.$$

**Exercise 4.6.6** Let $X$ be a simple graph with $n$ vertices and $e$ edges. If $\lambda$ is an eigenvalue of the adjacency matrix $A$ of $X$, show that $|\lambda| \leq \sqrt{\frac{2e(n-1)}{n}}$.

**Exercise 4.6.7** If two non-adjacent vertices of a graph $X$ are adjacent to the same set of vertices, show that 0 is an eigenvalue of its adjacency matrix.

**Exercise 4.6.8** The **eccentricity** $\mathrm{ecc}(x)$ of a vertex $x$ in a connected graph $X = (V, E)$ is the maximum of $d(x, y)$ as $v$ ranges over the vertices of $X$. **The radius** $\mathrm{rad}(X)$ of $X$ is defined as the minimum of $\mathrm{ecc}(x)$, $x \in V$. Prove that $\mathrm{rad}(X) \leq \mathrm{diam}(X) \leq 2\mathrm{rad}(X)$.

**Exercise 4.6.9** Calculate the eccentricity of every vertex of the path with $n$ vertices.

**Exercise 4.6.10** Let $X = (V, E)$ be a graph with $n$ vertices. If $X^c$ is the complement of $X$, show that the Laplacian of $X^c$ equals $n I_n - J_n - L(X)$. If the eigenvalues of $X$ are $0 = \mu_0 \leq \mu_1 \leq \cdots \leq \mu_{n-1}$, show that the eigenvalues of $X^c$ are 0 and $n - \mu_j$, for $1 \leq j \leq n - 1$.

**Exercise 4.6.11** Let $X = (V, E)$ be a graph. Show that the signless Laplacian matrix $Q$ of $X$ is a real and symmetric matrix and all its eigenvalues are non-negative. Prove that 0 is an eigenvalue of $Q$ if and only if $X$ is a bipartite graph.

**Exercise 4.6.12** Let $X = (V, E)$ be a graph. If $\lambda_1$ is the largest eigenvalue of its adjacency matrix, prove that

$$\overline{d} \leq \lambda_1 \leq \Delta,$$

where $\overline{d}$ is the average degree of $X$ and $\Delta$ is the maximum degree of $X$. When does equality occur?

**Exercise 4.6.13** Find an orthonormal basis of $\mathbb{R}^n$ formed by eigenvectors of the adjacency matrix of $K_n$.

**Exercise 4.6.14** Prove that $t^4 + t^3 + 2t^2 + t + 1$ cannot be the characteristic polynomial of an adjacency matrix of any graph.

**Exercise 4.6.15** Let $X$ be the graph obtained from the cycle on four vertices by adding one isolated vertex. Let $Y$ be the complete bipartite graph $K_{1,4}$. Prove that $X$ and $Y$ have the same eigenvalues, but are not isomorphic.

**Exercise 4.6.16** Let $X$ be a connected graph with $n$ vertices. If $A$ is the adjacency matrix of $X$, prove that $X$ is a regular graph if and only if there exists a polynomial $p$ with real coefficients such that $p(A) = J_n$, where $J_n$ is the all one $n \times n$ matrix.

**Exercise 4.6.17** The **odd girth** of a graph $X$ is the shortest length of an odd cycle. If $X$ and $Y$ have the same eigenvalues of their adjacency matrices, then prove that they have the same odd girth.

**Exercise 4.6.18** The **line graph** $\mathcal{L}(X)$ of a graph $X$ the edges of $X$ as vertices, two edges $e$ and $f$ of $X$ being adjacent in $\mathcal{L}(X)$ if they have common endpoint in $X$. Show that if $N$ is the incidence matrix of $X$, then the adjacency matrix of $\mathcal{L}(X)$ is $N^t N - 2I_m$, where $m$ is the number of edges of $X$.

**Exercise 4.6.19** Let $X$ be a $k$-regular graph. Show that if $\lambda$ is an eigenvalue of the adjacency matrix of $X$, then $k + \lambda - 2$ is an eigenvalue of the adjacency matrix of the line graph of $X$.

**Exercise 4.6.20** Prove that any eigenvalue of the adjacency matrix of a line graph is greater than or equal to $-2$.

# Chapter 5
# Trees

## 5.1 Forests, Trees, and Leaves

A **forest** is an acyclic graph, that is, a graph with no cycles. The connected components of a forest are called **trees**. Therefore, a **tree** is a connected graph that contains no cycles. In the figure below, we have a tree with seven vertices (Fig. 5.1).

Given a graph $X$ and a vertex $x$, we denote by $X - x$ the graph obtained by deleting the vertex $x$ and any edges incident with $x$. A **leaf** of $X$ is a vertex of degree one. We begin by proving the following result.

**Lemma 5.1.1** *Let $T$ be a tree with $n \geq 2$ vertices.*

*(1) The tree $T$ contains at least two leaves.*
*(2) For any leaf $x$, the graph $T - x$ is a tree with $n - 1$ vertices.*

***Proof*** For part (1), consider a maximal path $P$ in the tree $T$. Denote the endpoints of $P$ by $a$ and $b$. Every neighbour of $a$ or $b$ must be on the path $P$. Otherwise, this would violate the maximality of the path. However, if $a$ or $b$ had two neighbours in $P$, we would get a cycle, which is a contradiction with $T$ being a tree. Thus, $a$ and $b$ must be leaves.

For part (2), let $x$ be a leaf in $T$. We will show that $T - x$ is a tree. The graph $T - x$ is acyclic because deleting a vertex is not going to create any cycles. Given two distinct vertices $y$ and $z$ in $T - x$, consider a $(y, z)$-path joining them in $T$. Such a path exists because $T$ is connected. Any such path cannot involve $x$ for otherwise $x$ will have degree two or more. Therefore, a $(y, z)$-path in $T$ will exist in $T - x$. This shows that $T - x$ is connected and, therefore, a tree. ∎

We now give the following characterization of trees.

**Theorem 5.1.2** *Let $T = (V, E)$ be a graph with $n$ vertices. The following statements are equivalent.*

*(1) The graph $T$ is a tree.*

© Hindustan Book Agency 2022
S. M. Cioabă and M. R. Murty, *A First Course in Graph Theory and Combinatorics*,
Texts and Readings in Mathematics 55, https://doi.org/10.1007/978-981-19-0957-3_5

**Fig. 5.1**  A tree on seven
vertices

*(2)  The graph T is connected and has n − 1 edges.*
*(3)  The graph T has n − 1 edges and no cycles.*
*(4)  For any distinct vertices x, y ∈ V, there is a unique path joining them.*

***Proof***  To prove (1) implies (2), we use induction. By the previous lemma, let $x$
be a leaf of $T$ and consider the tree $T − x$ with $n − 1$ vertices. By the induction
hypothesis, it has $n − 2$ edges and together with the edge involving $x$, $T$ must have
$n − 1$ edges. The same argument shows that (1) implies (3).

To show that (2) implies (3), assume by contradiction that $T$ has a cycle. We
may delete edges from any cycle until we get a graph $T'$ which is acyclic and has $n$
vertices. The graph $T'$ is still connectedand, therefore, a tree. Hence, $T'$ has $n − 1$
edges. Thus, no edges were deleted from $T$ and $T = T'$ has no cycles.

We can also show that (3) implies (1) as follows. Let $T_1, \ldots, T_k$ be the connected
components of $T$, where $k \geq 1$. We have that $\sum_{j=1} |V(T_j)| = n$. As $T$ has no cycles,
each component $T_j$ is a tree and, therefore, $|E(T_j)| = |V(T_j)| − 1$ for $1 \leq j \leq k$.
Thus, the number of edges of $T$ is $n − k$. But as $T$ has $n − 1$ edges, this means that
$k = 1$ and so $T$ has only one connected component. Therefore, $T$ is a tree.

Finally, we must show the equivalence of (1) and (4). Clearly, (1) implies (4)
as otherwise $T$ would have a cycle. Conversely, if any two points have a unique
path joining them, there are no cycles in the graph; moreover, $T$ is connected. This
completes the proof.                                                                          ∎

## 5.2  Labelled Trees

Arthur Cayley (1821–1895) was an English mathematician who spent 14 years as a
lawyer, during which he published 250 mathematical papers. In total, he published
over 900 papers and notes covering almost every aspect of mathematics.

We consider the problem of counting trees on a given number of vertices. The
answer depends on whether we are counting unlabeled or labelled trees. For example,
when $n = 3$, there is exactly one unlabelled tree with $n$ vertices while there are exactly
three labelled trees. For the labelled trees situation, we consider trees whose vertex
set is $[n] = \{1, \ldots, n\}$. Two such labelled trees are considered the same if their edge
sets are equal (Fig. 5.2).

**Fig. 5.2** Unlabelled and labelled trees on three vertices

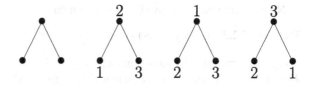

A classical result of Cayley states that the number of labelled trees on $n$ vertices is $n^{n-2}$. There are several proofs of this formula. We now use the principle of inclusion and exclusion and induction to deduce it.

For two non-negative integers $m$ and $n$, let $A$ be the set of all connected labelled graphs on vertex set $[n]$ having $m$ edges and define $G(n, m)$ as $|A|$. Let $F(n, m)$ denote the number of these graphs that have no leaves (vertices of degree one). For $1 \leq j \leq n$, let $A_j$ be the subset of $A$ consisting of those graphs where vertex $j$ has degree one. Thus, $F(n, m) = |A \setminus \cup_{j=1}^{n} A_j|$. Now observe that

$$|A_j| = G(n - 1, m - 1)(n - 1)$$

and, more generally,

$$|A_J| = G(n - |J|, m - |J|)(n - |J|)^{|J|},$$

for any $J \subseteq [n]$.

By the principle of inclusion and exclusion, we have that

$$F(n, m) = \sum_{J \subseteq [n]} (-1)^{|J|} G(n - |J|, m - |J|)(n - |J|)^{|J|}.$$

By collecting subsets of the same cardinality in the sum on the right, we obtain the following result.

**Theorem 5.2.1** *For any non-negative integers n and m,*

$$F(n, m) = \sum_{j=0}^{n} (-1)^j \binom{n}{j} G(n - j, m - j)(n - j)^j.$$

Theorem 5.1.2 tells us that any connected graph with $n$ vertices and $n - 1$ edges is necessarily a tree. Thus, $G(n, n - 1)$ is the number of labelled trees on $n$ vertices. Since every tree has a leaf, $F(n, n - 1) = 0$. We have the following corollary of the above result.

**Corollary 5.2.2** *If $T_n$ is the number of labelled trees on n vertices, then*

$$\sum_{j=0}^{n} (-1)^j \binom{n}{j} T_{n-j}(n - j)^j = 0.$$

Now we are ready to prove Cayley's formula.

**Theorem 5.2.3** (Cayley, 1889) *For $n \geq 2$, $T_n = n^{n-2}$.*

**Proof** We prove this by strong induction on $n$. For $n = 2$, the formula is clear as $T_2 = 1 = 2^{2-0}$. Let $n \geq 3$. Assume that $T_{n-j} = (n-j)^{n-j-2}$ for any $1 \leq j \leq n-2$. Plugging these values into the equation of Corollary 5.2.2, gives that

$$T_n + \sum_{j=1}^{n}(-1)^j \binom{n}{j}(n-j)^{n-2} = 0.$$

By Theorem 3.3.1, the latter sum is

$$\sum_{j=1}^{n}(-1)^j \binom{n}{j}(n-j)^{n-2} = -n^{n-2},$$

from which we deduce the theorem. ∎

The number $u_n$ of unlabelled trees on $n$ vertices does not have such a nice formula. However, one can prove that $2^n < u_n < 4^n$ for sufficiently large $n$ (see Exercise 5.6.5).

## 5.3 Spanning Trees

A **spanning subgraph** of a graph $X = (V, E)$ is a subgraph of $X$ whose vertex set is $V$. Informally, a spanning subgraph is obtained from $X$ by keeping the same vertex set and deleting some (possibly none) edges. A **spanning tree** of $X$ is a spanning subgraph which is a tree. We denote by $\tau(X)$ the number of spanning trees of $X$. Clearly, $X$ is connected if and only if $\tau(X) > 0$.

We will describe how we can calculate $\tau(X)$ using the eigenvalues of the Laplacian matrix described in the previous chapter. First, we describe a recurrence relation that is satisfied by $\tau(X)$. Given an edge $e$ of $X$, we denote by $X - e$ the graph obtained from $X$ by deleting the edge $e$. Note that the endpoints of $e$ still belong to $X - e$. It may happen that deleting $e$ increases the number of components of the graph, in which case we call $e$ a **cut edge** or a **bridge**. The **contraction** of $X$ by an edge $e$ with endpoints $x$ and $y$ is the graph obtained by replacing $x$ and $y$ by a single vertex whose incident edges are the edges other than $e$ that were incident to $u$ or $v$. The resulting multigraph, denoted $X/e$, has one less edge than $X$.

**Theorem 5.3.1** *Let $X$ be graph. If $e$ is an edge of $X$ that is not a loop, then*

$$\tau(X) = \tau(X - e) + \tau(X/e).$$

**Proof** The number of spanning trees $\tau(X)$ of $X$ equals the sum of the number of spanning trees that do contain $e$ and the number of spanning trees that contain $e$. The spanning trees of $X$ that omit $e$ are counted by $\tau(X - e)$. The family of spanning trees that contain $e$ is in bijective correspondence with the set of spanning trees of $X/e$. To see this, note that when we contract $e$ in a spanning tree that contains $e$, we obtain a spanning tree of $X/e$ because the resulting subgraph of $X/e$ is spanning, connected, and has the right number of edges. Since the other edges maintain their identity under contraction, no two trees are mapped to the same spanning tree of $X/e$ by this operation. Also, each spanning tree of $X$ arises in this way, and so the function is a bijection. This proves the theorem.                               ∎

The Laplacian of a graph $X = (V, E)$ is the matrix $L = D - A$, where $D$ is the diagonal degree matrix and $A$ is the adjacency matrix of $X$. Given a square $n \times n$ matrix $M$ with rows and columns indexed by $[n]$, denote by $M[i, j]$ the $(n - 1) \times (n - 1)$ matrix obtained by removing the $i$th row and the $j$th column of $M$. When $i = j$, we denote $M[i, i]$ by $M[i]$. **The cofactor matrix** $C$ is the $n \times n$ matrix defined as

$$C(i, j) = (-1)^{i+j} \det(M[i, j]),$$

for any $i, j \in [n]$. **The adjugate matrix** (or classical adjoint) $\mathrm{adj}(M)$ of $M$ is the transpose of its cofactor matrix $C$. From linear algebra, we know that

$$M\,\mathrm{adj}(M) = \mathrm{adj}(M)M = \det(M)I_n. \tag{5.3.1}$$

Gustav Robert Kirchhoff (1824–1887) was a German physicist, perhaps best known for the Kirchhoff's laws in electrical circuits. The following celebrated theorem of Kirchhoff from 1847 gives the number $\tau(X)$ via a determinant formula.

**Theorem 5.3.2** (Kirchhoff's Matrix-Tree Theorem) *Let $X$ be a graph with vertex set $[n]$ for some natural number $n \geq 1$ and Laplacian matrix $L$.*

*(1) For any $i \in [n]$, $\tau(X) = \det L[i]$.*
*(2) For any $i, j \in [n]$, $\tau(X) = (-1)^{i+j} \det L[i, j]$.*
*(3) $\mathrm{adj}(X) = \tau(X)J_n$.*
*(4) $\tau(X) = n^{-2} \det(J_n + L)$.*
*(5) If $0 = \mu_1 \leq \mu_2 \leq \cdots \leq \mu_n$ are the eigenvalues of $L$, then*

$$\tau(X) = \frac{\mu_2 \cdot \ldots \cdot \mu_n}{n}.$$

**Proof** Clearly, (2) is a more general case of (1), but we will prove (1) first and use it to prove (2). To prove (1), we use strong induction on the number of edges. We leave the base case details for the reader. For the induction step, we show that $\det L[i]$ satisfies the same recurrence relation as $\tau(X)$ in Theorem 5.3.1. Let $e = ij$ be an edge of $X$. Denote by $E$ the $([n] \setminus \{i\}) \times ([n] \setminus \{i\})$ matrix that has all entries zero except for $E(j, j) = 1$. Note that $L[i] = L(X - e)[i] + E$. This implies that

$$\det L[i] = \det L(X - e)[i] + \det L(X - e)[i, j]$$
$$= \det L(X - e)[i] + \det L[i, j],$$

where the second part follows since $L(X - e)[i, j] = L[i, j]$. When contracting the edge $e = ij$ of $X$ and creating $X/e$, assume that the vertex set of $X$ is $[n] \setminus \{i\}$. Note that $L(X/e)[j] = L[i, j]$. Using the previous equation, we get that

$$\det L[i] = \det L(X - e)[i] + \det L(X/e)[i].$$

From the induction hypothesis, we know that $\det L(X - e)[i] = \tau(X - e)$ and $\det L(X/e)[i] = \tau(X/e)$ and, therefore, by Theorem 5.3.1, we get that

$$\det L[i] = \tau(X - e) + \tau(X/e) = \tau(X),$$

which proves part (1).

For parts (2) and (3), note that $\mathrm{adj}(L)L = L\mathrm{adj}(L) = \det(L)I_n = 0$ since $L$ has 0 as an eigenvalue (see Theorem 4.5.2). This means that each column of $\mathrm{adj}(L)$ is contained in the eigenspace corresponding to the eigenvalue 0 of $L$. By part (3) of Theorem 4.5.2, we get that all the entries of any given column of $\mathrm{adj}(L)$ must be equal. Since the diagonal entries of $\mathrm{adj}(L)$ equal $\tau(X)$, it follows that all the entries of $\mathrm{adj}(L)$ equal $\tau(X)$.

For part (4), note first that $J_n^2 = nJ_n$ and $J_n L = L_n J = O_n$. Hence, $(nI_n - J_n)(J_n + L) = nL$ and, therefore,

$$\mathrm{adj}(nL) = \mathrm{adj}((nI_n - J_n)(J_n + L)) = \mathrm{adj}(J_n + L)\mathrm{adj}(nI_n - J_n).$$

But $\mathrm{adj}(nI_n - J_n) = n^{n-2}J_n$ by Cayley's formula and $\mathrm{adj}(nL) = n^{n-1}\mathrm{adj}(L)$. We therefore deduce that $n\mathrm{adj}(L) = [\mathrm{adj}(J_n + L)]J_n$. By part (3), $\mathrm{adj}(L) = \tau(X)J_n$ and so we obtain that $n\tau(X)J_n = [\mathrm{adj}(J + L)]J_n$. Multiplying both sides of the equation by $(J + L)$ on the left gives that $n\tau(X)(J + L)J = [\det(J + L)]J$. But $(J_n + L)J_n = J_n^2 + LJ_n = nJ_n$. Putting everything together, we get the desired result.

For part (5), note that the eigenvalues of $J_n + L$ are $n, \mu_2, \ldots, \mu_n$, and $\det(J_n + L) = n \prod_{j=2}^{n} \mu_j$. The result follows using part (4).  ∎

**Example 5.3.3** Cayley's theorem can be deduced easily from this more general result, as follows. The number of trees on a vertex set $[n]$ is the number of spanning trees of the complete graph $K_n$. The adjacency matrix of $K_n$ is $J_n - I_n$, where $J_n$ is the all one $n \times n$ matrix and $I_n$ is the identity matrix. The Laplacian matrix of the complete graph is $L(K_n) = (n - 1)I_n - (J_n - I_n)$. Any diagonal cofactor is

$$\det[(n - 1)I_{n-1} - (J_{n-1} - I_{n-1})].$$

By Example 4.2.3 in Chap. 4, we see that this determinant is the characteristic polynomial of the graph $K_{n-1}$ evaluated at $t = n - 1$ which is $[t - (n - 2)](t + 1)^{n-2} = n^{n-2}$.

In the case that $X$ is a $k$-regular graph, we can derive a formula for $\tau(X)$ using the eigenvalues of the adjacency matrix.

**Corollary 5.3.4** *Let $X$ be a $k$-regular graph. If $k = \lambda_1 \geq \lambda_2 \geq \cdots \geq \lambda_n$ are the eigenvalues of its adjacency matrix, then*

$$\tau(X) = n^{-1} \prod_{j=2}^{n} (k - \lambda_j).$$

**Example 5.3.5** The number of spanning trees of the bipartite graph $K_{n,n}$ is $n^{2n-2}$.


## 5.4   MST, BFS, and DFS

In many contexts in which graph theory is applied, we consider **weighted graphs**. That is, we suppose we have a graph $X = (V, E)$ together with a *weight* function $w : E \rightarrow \mathbb{R}^+$. For example, our graph could be a network of cities, and the weight function could be the cost of putting a communication network between the two cities. We will be interested in finding a subgraph so that the graph is connected and the total *cost*, which is the sum of the weights of the edges in the subgraph, is minimal. Clearly, if there is a cycle, we can delete a *costly* edge and so, what we are searching is a spanning tree whose *cost* is minimal. We call such a tree a **minimum spanning tree**. Of course, it need not be unique.

There is a fundamental algorithm, called **Kruskal's algorithm**, which determines a minimum spanning tree of any connected graph in the most *naive* fashion. The algorithm was discovered in 1956 by the American mathematician Joseph Kruskal (1928–2010) and it can be described as follows. Choose edge $e_1$ with $w(e_1)$ minimal. Eliminate it from the list. Inductively choose $e_2, \ldots, e_{n-1}$ in the same manner, subject to the constraint that we have no cycle. The required spanning tree is the subgraph with these edges. Before we prove that this **greedy algorithm**, actually works, we illustrate this with an example.

Consider the weighted adjacency matrix, giving the cost of building a road from one city to another. An infinite entry indicates there is a mountain in the way and a road cannot be built. The question is to determine the least cost of making all the cities reachable from each other. This amounts to finding a spanning tree with minimum *cost*.

**Example 5.4.1** The algorithm proceeds first by finding an edge of minimum weight, $AB$ say. It then deletes the entry. Next, find the next smallest entry, $BC$ say. Continue in this way, and whenever an edge is chosen which produces a cycle, do not select it.

Thus, in the example below, $AC$ is the next smallest entry, but we would not choose it for it produces a cycle. Thus, the next entry to choose is $DE$ followed by $BE$.

$$
\begin{array}{c@{\quad}ccccc}
 & A & B & C & D & E \\
\begin{array}{c} A \\ B \\ C \\ D \\ E \end{array} &
\left[\begin{array}{ccccc}
0 & 3 & 5 & 11 & 9 \\
3 & 0 & 3 & 9 & 8 \\
5 & 3 & 0 & \infty & 10 \\
11 & 9 & \infty & 0 & 7 \\
9 & 8 & 10 & 7 & 0
\end{array}\right]
\end{array}
$$

**Theorem 5.4.2** *In a weighted connected graph X, Kruskal's algorithm constructs a minimum weight spanning tree.*

*Proof* Clearly, the algorithm produces a tree since an edge which produces a cycle is never selected. Let $T$ be the tree produced by the algorithm. Let $T^*$ be a minimum weight spanning tree. If $T = T^*$, we are done. If not, let $e$ be the first edge chosen for $T$ but not in $T^*$. Let $T^* + e$ be the graph obtained from $T^*$ by adding the edge $e$. This graph contains a cycle $C$. Since $T$ had no cycle, $C$ has an edge $e' \notin E(T)$. Now consider the spanning tree $(T^* + e) - e'$. Since $T^*$ contains $e'$ and all the edges of $T$ chosen before $e$, both $e'$ and $e$ are available when the algorithm chooses $e$ and, therefore, $w(e) \leq w(e')$. Thus, $(T^* + e) - e'$ is a spanning tree with weight at most that of $T^*$ that agrees with $T$ for a longer initial list than $T^*$ does. Repeating this process eventually produces a minimum spanning tree that agrees with $T$. ∎

The number of spanning trees of a connected graph $X$ can be quite large and in many situations, we would like to construct a spanning with certain properties. In this section, we describe two of the most commonly used spanning trees, called breadth-first search or BFS spanning tree and depth-first search or DFS spanning tree. We conclude the section by discussing another famous algorithm, Kruskal's algorithm for constructing a minimum weight spanning tree of a weighted graph.

To discuss BFS and DFS trees, we will need the notion of a rooted tree.

**Definition 5.4.3** A **rooted tree** is a pair $(T, r)$ where $T$ is a tree and $r$ is a vertex of $T$ called its **root**.

Note that $T$ may be labelled or unlabelled.

**Definition 5.4.4** Let $(T, r)$ be a rooted tree. The **level** $\ell(x)$ of a vertex $x$ of $(T, r)$ is defined as the distance from $x$ to the root $r$.

In some situations, it is convenient to draw a rooted tree by placing the root at the top and the rest of the neighbours below the root, with their position depending on their level (closer vertices to the root higher on the page). See Fig. 5.3 for some examples. For the rooted tree on the left, $\ell(a) = 0$ as $a$ is the root and $\ell(x) = 1$ for any other vertex. For the rooted tree on the right, $\ell(c) = 0$ as $c$ is the root, $\ell(a) = 1$ and $\ell(b) = \ell(d) = 2$.

To understand the intuition behind the BFS and DFS trees, we first describe the BFS and DFS traversal sequence of a given rooted tree. For a rooted tree $(T, r)$, a

**Fig. 5.3** Same labelled tree
rooted at different vertices

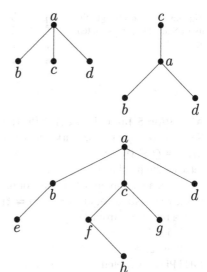

**Fig. 5.4** A rooted tree

breadth-first search or BSF traversal sequence of it is an ordered list of its vertices
essentially according to their levels. For the tree in Fig. 5.4, a BFS traversal sequence
would be $(a, b, c, d, e, f, g, h)$. A depth-first search or DFS traversal of $(T, r)$ is an
ordered list of its vertices that is best explained graphically using Fig. 5.4. Imagine
the edges of the tree as forming a fence and starting at the root $a$ and keeping the fence
on the left until we return at $a$. If we write down the vertices of $(T, r)$ when we first
encounter them during this search, we obtain the sequence $(a, b, e, c, f, h, g, d)$.

We leave the reader to figure out the traversal sequences for the trees in Fig. 5.3.
In order to describe how to construct BFS and DFS spanning trees, we need the
following useful notion related to rooted trees.

**Definition 5.4.5** Let $(T, r)$ be a rooted tree. For a vertex $x$ that is not the root $r$, the
**parent** $p(x)$ of $x$ is the neighbour of $x$ on the unique path joining $x$ to the root $r$. If
$y$ is the parent of $x$, we say that $x$ is the **child** of $x$.

The edge set of a rooted tree $(T, r)$ consists of all the edges of the form $\{x, p(x)\}$,
where $x$ ranges over all non-root vertices of $(T, r)$.

To describe the BFS tree, we will need a data structure called a queue. Think of a
socially distanced queue waiting for your favourite patio to open. This is an ordered
list $Q$ of vertices which is updated by either adding a new element to one end (the
tail of $Q$) or removing an element on the other end (the head of $Q$). For example, if
$Q = (a, b, c)$, the head of $Q$ is $a$ and its tail is $c$. If we add an element $d$ to $Q$, then $Q$
becomes $(a, b, c, d)$. If now we remove the head of $Q$, then $Q$ becomes $(b, c, d)$. In
computer science, a queue is also a type of data structure abbreviated FIFO, meaning
*First In, First Out*.

Given a connected graph $X = (V, E)$ and a vertex $r$ of it, we will construct the
breadth-first search BFS tree $(T, r)$ of $X$ as follows.

**Fig. 5.5** A connected graph, its BFS and DFS trees rooted at $a$

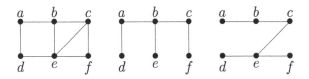

**Algorithm 5.4.6** (BFS tree) INPUT: $X = (V, E)$ connected graph, $r \in V$.
Set $Q = (r)$; $\ell(r) = 0$; mark $r$ as visited
while $Q \neq \emptyset$
let $x$ be the head of $Q$
if $x$ has a neighbour $y$ that has not been visited, then
mark $y$ as visited; $p(y) = x$; $\ell(y) = \ell(x) + 1$; add $y$ to $Q$
else remove $x$ from $Q$
end if
end while
OUTPUT: A rooted tree $(T, r)$ with a parent function $p$ and a level function $\ell$.

We describe how the BFS algorithm works on the example in Fig. 5.5 with the root $r$ being the vertex $a$. Note that when we have to choose from the neighbours of a given vertex, we use lexicographic order. The behaviour of the queue $Q$ as the algorithm progresses is the following:

$$(a) \to (a, b) \to (a, b, d) \to (b, d) \to (b, d, c) \to (b, d, c, e)$$
$$\to (d, c, e) \to (c, e) \to (c, e, f) \to (e, f) \to (f) \to \emptyset.$$

We summarize below the properties of the BFS tree. They can be proved via induction on the number of vertices.

**Proposition 5.4.7** *Let $X = (V, E)$ be a connected graph and $r \in V$. Let $(T, r)$ be the rooted BFS tree of $X$.*

*(1)  The level function gives the distance from the root $r$ in $X$ and $\ell(x)$ equals the distance between $x$ and $r$ in $X$.*
*(2)  Any edge of $X$ connects two vertices whose levels differ by at most 1.*

Another application of the DFS algorithm is that it can be used to detect if a graph is bipartite by either finding a partition of its vertex set into two independent sets (if it is) or by finding the shortest odd cycle (if the graph is not bipartite). One could start with an arbitrary root $r$ and construct the BFS tree $(T, r)$ rooted at $r$. For any $\ell \geq 1$, one would check if there are any edges in $X$ joining two vertices at the same level $\ell$ in $T$. The smallest $\ell$ where this happens will indicate the existence of an odd cycle of length $2\ell + 1$. If there are no such edges, then there are no odd cycles, and we can partition the vertices of $X$ according to the parity of their levels in $T$: the vertices of even level (including $r$) in one part and the vertices of odd level in the other part. Each of these parts is an independent set, and this argument shows that $X$ is bipartite in this situation.

To describe the DFS tree, we will need a data structure called a stack. This is an ordered list $S$ of vertices which is updated at one end only (the top of the stack or left endpoint in our horizontal notation). Think of a stack of plates. For example, if $S = (a, b, c)$, the top of $S$ is $a$. If we add an element $d$ to $S$, then $S$ becomes $(d, a, b, c)$. If now we remove the top of $S$, then $S$ becomes $(a, b, c)$. In computer science, a stack is also a type of data structure abbreviated LIFO, meaning *Last In, First Out*.

Given a connected graph $X = (V, E)$ and a vertex $r$ of it, we will construct the breadth-first search BFS tree $(T, r)$ of $X$ as follows.

**Algorithm 5.4.8** (DFS tree) INPUT: $X = (V, E)$ connected graph, $r \in V$.
Set $S = (r)$; mark $r$ as visited
while $S \neq \emptyset$
let $x$ be the top of $S$
if $x$ has a neighbour $y$ that has not been visited, then
mark $y$ as visited; $p(y) = x$; add $y$ to $S$
else remove $x$ from $S$
end if
end while
OUTPUT: A rooted tree $(T, r)$ with a parent function $p$.

We describe how the BFS algorithm works on the example in Fig. 5.5 with the root $r$ being the vertex $a$. Note that when we have to choose from the neighbours of a given vertex, we use lexicographic order. The behaviour of the stack $S$ as the algorithm progresses is the following:

$$(a) \to (b, a) \to (c, b, a) \to (e, c, b, a) \to (d, e, c, b, a) \to (e, c, b, a)$$
$$\to (f, e, c, b, a) \to (e, c, b, a) \to (c, b, a) \to (b, a) \to (a) \to \emptyset.$$

**Definition 5.4.9** Let $(T, r)$ be a rooted tree. A vertex $x$ of $T$ is called an **ancestor** of a vertex $y$ of $T$ if $x$ is contained in the unique path between the root $r$ and $y$. If this happens, the vertex $y$ is called a **descendant** of $x$. Two vertices of $T$ are called **related** if one is an ancestor of the other.

We summarize below the properties of the DFS tree. We leave the proof of this result as an exercise.

**Proposition 5.4.10** *Let $X = (V, E)$ be a connected graph and $r \in V$. Let $(T, r)$ be the rooted DFS tree of $X$.*

*(1) The endpoints of any edge of $X$ are related in $T$.*
*(2) The deletion of $r$ disconnects the graph $X$ if and only if $r$ has at least two children in $T$.*

## 5.5   Lagrange's Inversion Formula

There is an alternate method of deriving Cayley's formula due to the Hungarian-American mathematician George Pólya (1887–1985) whose work on enumeration we will encounter again in Sect. 7.5. His method uses generating functions and an important idea called the Lagrange inversion formula, which we describe below.

We begin with a simple counting principle. For $n \geq 0$, let $g_n$ denote the number of graphs on $n$ vertices for which each connected component enjoys a certain property $P$. Denote by $c_n$ the number of connected graphs with property $P$. We can define two (exponential) generating functions:

$$G(x) := \sum_{n=0}^{\infty} \frac{g_n x^n}{n!}, \quad C(x) := \sum_{n=0}^{\infty} \frac{c_n x^n}{n!}.$$

As the number of $k$ component graphs whose connected components have property $P$ has generating function

$$\frac{C(x)^k}{k!}$$

we see that

$$G(x) = e^{C(x)} - 1.$$

If we let $T_n$ be the number of trees on $n$ labelled vertices, then the generating function for the number of rooted trees is clearly

$$Y(x) = \sum_{n=0}^{\infty} \frac{n T_n x^n}{n!},$$

as there are $n$ choices for the roots. Thus, $Y^k/k!$ is the generating function for forests of $k$ rooted trees. If we join each root node of a forest of rooted trees to a new node, we obtain a rooted tree with one more node than the original forest. Every rooted tree (or at least those with more than one node) can be obtained uniquely this way. Thus,

$$Y(x) = x + xY(x) + x\frac{Y(x)^2}{2!} + \cdots$$

which gives

$$x = Y(x)e^{-Y(x)}. \tag{5.5.1}$$

We want to "solve" for $Y(x)$ in terms of $x$ and this is precisely what the Lagrange inversion formula will do.

Given a formal power series

$$f(t) = \sum_{n=-\infty}^{\infty} c_n t^n,$$

we use the notation

$$[t^n](f(t)) := c_n.$$

Thus, for instance, $[t^n](f(t)) = [t^{n+m}](t^m f(t))$, a fact we will use later in the proof of the proposition below.

**Theorem 5.5.1** (Lagrange inversion formula) *Suppose $f(z)$ is analytic in a neighbourhood of $z = 0$ with $f(0) = 0$ and $f'(0) \neq 0$. Then $f^{-1}$ is analytic in a neighbourhood of $z = 0$ and*

$$[z^n](f^{-1}(z)) = [z^{n-1}]\left(\frac{z^n}{n f(z)^n}\right).$$

*Proof* Since $f'(0) \neq 0$, we have by the inverse function theorem (see p. 62 of Ram Murty 2020) that $f^{-1}(z)$ is well-defined and analytic in a neighbourhood of $f(0) = 0$. Consequently, it has a power series expansion in a neighbourhood of zero. By the Cauchy residue theorem,

$$[z^n](f^{-1}(z)) = \frac{1}{2\pi i} \int_C \frac{f^{-1}(w) dw}{w^{n+1}},$$

with $C$ being a sufficiently small circle centred at zero. We change variables in the integral by setting $w = f(v)$ which is a conformal map if $C$ is of sufficiently small radius. Since $f^{-1}(f(v)) = v$, we have

$$[z^n](f^{-1}(z)) = \frac{1}{2\pi i} \int_{C'} \frac{v f'(v) dv}{f(v)^{n+1}},$$

where $C'$ is the closed contour image of $C$ under our conformal mapping. Our integral can be re-written as

$$-\frac{1}{2\pi i n} \int_{C'} v\, d\left(\frac{1}{f(v)^n}\right) = \frac{1}{2\pi i n} \int_{C'} \left(\frac{1}{f(v)^n}\right) dv,$$

on integrating by parts and noting that the residue of $d(v/f(v)^n)$ at $v = 0$ is zero. Thus, by the Cauchy residue theorem, we conclude that

$$[z^n](f^{-1}(z)) = [z^{-1}]\left(\frac{1}{n f(z)^n}\right) = [z^{n-1}]\left(\frac{z^n}{n f(z)^n}\right),$$

the last equality being clear by shifting the power series appropriately. ∎

Applying the Lagrange inversion formula (exercise) to Eq. (5.5.1) gives Cayley's formula.

## 5.6  Exercises

**Exercise 5.6.1** Prove that in any tree with two or more vertices, every edge is a bridge. Show that if $T$ is a tree on $n \geq 2$ vertices, then for any edge $e$ of $T$, $T - e$ is a forest of two trees.

**Exercise 5.6.2** Let $X$ be a connected graph with $n$ vertices. Show that $X$ has exactly one cycle if and only if $X$ has $n$ edges. Prove that a graph with $n$ vertices and $e$ edges contains at least $e - n + 1$ cycles.

**Exercise 5.6.3** Let $d_1 \geq \cdots \geq d_n$ be natural numbers.

(1) Show that there exists a tree with vertex set $[n]$ where vertex $j$ has degree $d_j$ for $1 \leq j \leq n$ if and only if

$$d_1 + \cdots + d_n = 2n - 2.$$

(2) If $d_1 + \cdots + d_n = 2n - 2$, then show that the number of trees in part (1) is $\frac{(n-2)!}{(d_1-1)!\ldots(d_n-1)!}$.

(3) Use part (2), to give another proof of Cayley's theorem that there are $n^{n-2}$ labelled trees on vertex set $[n]$.

**Exercise 5.6.4** Let $T$ be a tree with $n \geq 2$ vertices. Show that for any vertex $x$ of $T$ of degree $d(x)$, the graph $T - x$ is a forest with exactly $d(x)$ trees.

**Exercise 5.6.5** For $n \geq 2$, let $u_n$ denote the number of unlabelled trees on $n$ vertices.

(1) For $3 \leq n \leq 6$, determine $u_n$ exactly.

(2) Prove that for any $n \geq 2$, $\frac{n^{n-2}}{n!} < u_n < \binom{2n-2}{n-1} < 4^n$.

(3) Show that for $n$ sufficiently large, $2^n < u_n$.

**Exercise 5.6.6** Let $T$ and $T'$ be two distinct trees on the same vertex set $V$.

(1) Show that for each edge $e \in E(T) \setminus E(T')$, there exists $e' \in E(T') \setminus E(T)$ such that $(T \setminus \{e\}) \cup \{e'\}$ is a tree.

(2) Show that for each edge $f' \in E(T') \setminus E(T)$, there exists $f \in E(T) \setminus E(T')$ such that $(T \cup \{f'\}) \setminus \{f\}$ is a tree.

**Exercise 5.6.7** Let $T_n$ be the number of labelled trees with $n$ vertices. Prove that

$$2(n - 1)T_n = \sum_{i=1}^{n-1} \binom{n}{i} T_i T_{n-i} i (n - i).$$

**Exercise 5.6.8** Let $T$ be a labelled tree on vertex set $[n]$, where $n \geq 3$. Construct the following word $(a_1, \ldots, a_{n-2})$ of length $n - 2$ whose entries belong to the set $[n]$. Let $b_1$ be the smallest leaf and denote by $a_1$ its neighbour. Let $T_1 = T$ and $T_2 = T_1 - b_1$. Repeat the procedure: let $b_2$ be the smallest leaf in $T_2$ and let $a_2$ be its neighbour. For $1 \leq j \leq n - 1$, let $b_j$ be the smallest leaf in $T_j$ and denote by $a_j$ its neighbour. The sequence $(a_1, \ldots, a_{n-2})$ is called the Prüfer code of the tree $T$.

(1) Determine the Prüfer code of every labelled tree on 3 and 4 vertices.
(2) Construct the tree whose Prüfer code is $(1, 1, 2, 2, 7, 8)$.
(3) Show that $a_{n-1} = n$ and that the edge set of $T$ consists of the pairs $\{a_j, b_j\}$, $1 \leq j \leq n - 1$.
(4) For $x \in [n]$, show that the number of occurrences of $x$ among $a_1, \ldots, a_{n-2}$ equals $d_T(x) - 1$, where $d_T(x)$ is the degree of $x$ in $T$.
(5) Show that $b_k$ is the smallest element not contained in the set $\{a_1, \ldots, a_{k-1}\} \cup \{b_k, \ldots, b_{n-1}\}$.
(6) Show that the correspondence $T \mapsto (a_1, \ldots, a_{n-2})$ is a bijection and give another proof of Cayley's theorem.

**Exercise 5.6.9** Using Prüfer codes, give another proof of part (2) of Exercise 5.6.3.

**Exercise 5.6.10** Let $G(r, s; m)$ be the number of connected bipartite graphs with partite sets of size $r$ and $s$ and with $m$ edges, and let $F(r, s; m)$ be the number of these whose vertex degrees are each 2 or more. Prove that

$$F(r, s; m) = \sum_{i,j} \binom{r}{i}\binom{s}{j}(-1)^{i+j}G(r - i, s - j; m - i - j)(s - j)^i(r - i)^j.$$

**Exercise 5.6.11** Putting $m = r + s - 1$ in the previous exercise, notice that $G(r, s; r + s - 1)$ counts the number $T(r, s)$ (say) of spanning trees in the bipartite graph $K_{r,s}$. Deduce that

$$0 = \sum_{i,j} \binom{r}{i}\binom{s}{j}(-1)^{i+j}T(r - i, s - j)(s - j)^i(r - i)^j$$

and that $T(r, s) = r^{s-1}s^{r-1}$.

**Exercise 5.6.12** The **Wiener index** of a graph $X = (V, E)$ is $W(X) = \sum_{x,y \in V} d(x, y)$, where $d(x, y)$ denotes the distance from $x$ to $y$. Show that if $X$ is a tree with $n$ vertices, then
$$W(K_{1,n-1}) \leq W(X) \leq W(P_n).$$

**Exercise 5.6.13** A communication link is desired between five universities in Canada: Queen's, Toronto, Waterloo, McGill and UBC. With obvious notation, the matrix below gives the cost (in thousands of dollars) of building such a connection between any two of the universities.

$$
\begin{array}{c c c c c c}
 & Q & T & W & M & U \\
Q & - & 350 & 400 & 300 & 1200 \\
T & 350 & - & 100 & 600 & 1300 \\
W & 400 & 100 & - & 700 & 1400 \\
M & 300 & 600 & 700 & - & 1600 \\
U & 1200 & 1300 & 1400 & 1600 & - 
\end{array}
$$

Use the greedy algorithm to determine the minimal cost so that all universities are connected.

**Exercise 5.6.14** Every tree with maximum degree $d \geq 2$ has at least $d$ leaves. For any $n > d \geq 2$, construct a tree with $n$ vertices and maximum degree $d$ for each $n > d \geq 2$.

**Exercise 5.6.15** Let $X$ be a graph with $n \geq 3$ vertices such that by deleting any vertex of $X$, we obtain a tree. Find $X$.

**Exercise 5.6.16** Show that every connected graph $X$ contains at least two vertices $x$ with the property that $X - x$ is connected. What are the trees on $n$ vertices that contain exactly two vertices with this property?

**Exercise 5.6.17** Show that the graph obtained from $K_n$ by removing one edge has $(n-2)n^{n-3}$ spanning trees.

**Exercise 5.6.18** Let $G_n$ be the graph obtained from the path $P_n$ by adding one vertex adjacent to all the vertices of the path $P_n$. Determine the number of spanning trees of $G_n$.

**Exercise 5.6.19** If $G$ is a graph on $n$ vertices having maximum degree $\Delta \geq 2$ and diameter $D \geq 1$. Show that

$$
n \leq \begin{cases} 2D + 1, & \text{if } \Delta = 2, \\ \frac{\Delta[(\Delta-1)^D + 1]}{\Delta - 2} + 1, & \text{if } \Delta \geq 3. \end{cases}
$$

**Exercise 5.6.20** The **centre** of a graph $X$ is the subgraph induced by the vertices of minimum eccentricity. Show that the centre of a tree is a vertex or an edge.

# Reference

M. Ram Murty, *A Second Course in Analysis*. IMSC Lecture Notes 4 (Hindustan Book Agency, 2020)

# Chapter 6
# Möbius Inversion and Graph Colouring

## 6.1 Posets and Möbius Functions

August Ferdinand Möbius (1790–1868) was a German mathematician and astronomer who introduced the function which bears his name in 1831 and proved the well-known inversion formula. Möbius was an assistant to Carl Friedrich Gauss (1777–1855) and made important contributions in geometry and topology. That this function can be generalized to study various combinatorial structures was a brilliant idea of the Italian-American mathematician Gian-Carlo Rota (1932–1999). His insight that the Möbius function should be viewed as a function of two variables on a partially ordered set transformed combinatorics, and embodied a fundamental counting principle. This generalized Möbius function allows one to translate many combinatorial problems into problems about the location of zeroes of certain polynomials. This Möbius function is now an important tool in algebra, combinatorics, and number theory.

A **partially ordered set** or **poset** is a pair $(P, \leq)$, where $P$ a set and a binary relation $\leq$ on $P$ satisfying the following properties:

(1) $x \leq x$ for all $x \in P$ (reflexive property);
(2) $x \leq y$ and $y \leq x$ imply $x = y$ (antisymmetric property); and
(3) $x \leq y$, $y \leq z$ implies $x \leq z$ (transitive property).

We call $\leq$ a **partial order** on $P$. If $x \leq y$ and $x \neq y$, we write that $x < y$. An **interval** $[x, y]$ consists of elements of $z \in P$ satisfying $x \leq z \leq y$. A poset $P$ is called **locally finite** if every interval is finite. We say $y$ **covers** $x$ if $x \leq y$ and the interval $[x, y]$ consists of only two elements, namely, $x$ and $y$. The **Hasse diagram** of $(P, \leq)$ is given by representing elements of $P$ as points in the Euclidean plane, joining $x$ and $y$ by a line whenever $y$ covers $x$ and putting $y$ *higher* than $x$ on the plane. Here are some examples of posets (Figs. 6.1 and 6.2).

**Example 6.1.1** Let $S$ be a finite set. The collection $\mathcal{P}(S)$ of all subsets of $S$, partially ordered by set inclusion, is a finite poset.

© Hindustan Book Agency 2022
S. M. Cioabă and M. R. Murty, *A First Course in Graph Theory and Combinatorics*,
Texts and Readings in Mathematics 55, https://doi.org/10.1007/978-981-19-0957-3_6

**Fig. 6.1** The Hasse diagram
of $\mathcal{P}([3])$

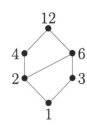

**Fig. 6.2** The poset of
divisors of 12

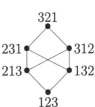

**Fig. 6.3** The poset $S_3$ with
the Bruhat order

**Example 6.1.2** Let $n$ be a natural number. The set $D(n)$ of the natural number divisors of $n$ with the binary relation of integer division forms a poset.

**Example 6.1.3** Let $\mathbb{N}$ be the set of natural numbers. We can define a partial order on this by divisibility. While $\mathbb{N}$ is infinite, this is a locally finite poset.

**Example 6.1.4** Let $S$ be a finite set. Given two partitions $\alpha$ and $\beta$ we say $\alpha \leq \beta$, if every block of $\beta$ is contained in a block of $\alpha$. We call $\beta$ a **refinement** of $\alpha$. The collection of partitions of $S$, $\Pi(S)$, with the above binary relation, is a poset.

**Example 6.1.5** We can define a partial order on the elements of the symmetric groups $S_n$, as follows. Let $\sigma, \tau \in S_n$. The permutation $\tau \in S_n$ is said to be a **reduction** of $\sigma$ if there exist $i, j \in [n]$ such that $i < j, \sigma(i) > \sigma(j)$ and $\tau(k) = \sigma(k)$ for all $k \neq i, j$. Clearly, $\tau(i) = \sigma(j) < \sigma(i) = \tau(j)$. Define a binary relation on $S_n$ as follows: $\eta \leq \pi$ if we can obtain $\eta$ by a sequence of reductions from $\pi$. This is called the **Bruhat order** on the symmetric group, which makes $S_n$ into a poset.

**Example 6.1.6** Let $q$ be a prime power and $n$ a natural number. Denote by $V(n, q)$ the $n$-dimensional vector space over the finite field of $q$ elements. The collection of its subspaces partially ordered by inclusion forms a finite poset.

Given two posets $(P_1, \leq_1)$ and $(P_2, \leq_2)$, we can define their **direct product** as $(P_1 \times P_2, \leq)$, with partial order

$$(x_1, y_1) \leq (x_2, y_2) \quad \text{iff} \quad x_1 \leq_1 x_2 \quad \text{and} \quad y_1 \leq_2 y_2.$$

Let $P$ be a poset and denoted by $I(P)$ the set of intervals of $P$. Let $\mathbb{F}$ be a field. For a function $f : I(P) \to \mathbb{F}$ and $[x, y] \in I(P)$, we will write $f(x, y)$ for $f([x, y])$. The **incidence algebra** $I(P, \mathbb{F})$ of the poset $P$ over the field $\mathbb{F}$ is the $\mathbb{F}$-algebra of functions $f : I(P) \to \mathbb{F}$, where the multiplication of two functions $f, g : I(P) \to \mathbb{F}$ is defined as $fg : I(P) \to \mathbb{F}$, where

$$(fg)(x, y) = \sum_{x \leq z \leq y} f(x, z)g(z, y), \forall x \leq y.$$

Given a locally finite poset $P$, its **Möbius function** $\mu$ is a map

$$\mu : P \times P \to \mathbb{Z},$$

defined recursively as follows. For $x, y \in P$, the Kronecker symbol $\delta(x, y)$ equals 1 if $x = y$ and 0, otherwise. If $x \leq y$, define $\mu(x, y)$ by the recursion

$$\sum_{z : x \leq z \leq y} \mu(x, z) = \delta(x, y). \tag{6.1.1}$$

Otherwise, $\mu(x, y)$ is defined as 0.

Observe that this equation can be written in *matrix form* as follows. Define the **zeta function** of $P$ by $\zeta(x, y) = 1$ if $x \leq y$ and zero otherwise. If for the moment, assume $P$ is finite and list its elements in some order: $z_1, \ldots, z_n$. If $Z$ is the matrix whose $(i, j)$th entry is $\zeta(z_i, z_j)$ and $M$ is the matrix whose $(i, j)$th entry is $\mu(z_i, z_j)$, then $MZ = I_n$. This follows from Eq. (6.1.1). Thus, $M$ is the inverse of the matrix $Z$. Since the inverse is both a left inverse and a right inverse, we deduce that $ZM = I_n$ which means that

$$\sum_{x \leq z \leq y} \mu(z, y) = \delta(x, y), \forall x \leq y \leq P. \tag{6.1.2}$$

**Theorem 6.1.7** (Möbius Inversion for Posets, version 1) *Let $(P, \leq)$ be a locally finite poset and $\mathbb{F}$ a field. Let $f, g : P \to \mathbb{F}$ be two functions. The following statements are equivalent:*

*(1)* $f(x) = \sum_{y : y \leq x} g(y), \quad \forall x \in P.$
*(2)* $g(x) = \sum_{y : y \leq x} \mu(y, x) f(y), \quad \forall x \in P.$

**Proof** For the direct implication, let $x \in P$. By changing the order of summation and using Eq. (6.1.2), we have that

$$\sum_{y:y\leq x} \mu(y,x)f(y) = \sum_{y:y\leq x} \mu(y,x) \sum_{z:z\leq y} g(z)$$

$$= \sum_{z:z\leq x} g(z) \sum_{y:z\leq y\leq x} \mu(y,x)$$

$$= \sum_{z:z\leq x} g(z)\delta(z,x)$$

$$= g(x)$$

as required.

For the converse implication, let $x \in P$. By changing the order of summation and Eq. (6.1.1), we obtain that

$$\sum_{y:y\leq x} g(y) = \sum_{y:y\leq x} \sum_{z:z\leq y} \mu(z,y)f(z)$$

$$= \sum_{z:z\leq x} f(z) \sum_{y:z\leq y\leq x} \mu(z,y)$$

$$= \sum_{z:z\leq x} f(z)\delta(z,x)$$

$$= f(x),$$

as required.                                                                                ∎

The following result can be proved by a similar argument, and its proof is left as an exercise.

**Theorem 6.1.8** (Möbius Inversion for Posets, version 2) *Let $(P, \leq)$ be a locally finite poset and $\mathbb{F}$ a field. Let $f, g : P \to \mathbb{F}$ be two functions. The following statements are equivalent:*

(1)  $f(x) = \sum_{y:y\geq x} g(y), \forall x \in P.$
(2)  $g(x) = \sum_{y:y\geq x} \mu(x,y)f(y), \forall x \in P.$

## 6.2  Applications of Möbius Inversion

Given a poset $(P, \leq)$ and $x, y \in P$, we say that $z$ is a **lower bound** of $x$ and $y$ if $z \leq x$ and $z \leq y$. Any maximal element of the set of lower bounds for $x$ and $y$ is called **a greatest lower bound**. Such elements need not be unique, as simple examples can show. Such an example is the poset $S_3$ with the Bruhat order in Fig. 6.3. The notions of **upper bound** and **least upper bound** are similarly defined.

A **lattice** $L$ is a poset $(L, \leq)$ with the property that any two elements have a unique greatest lower bound and a unique least upper bound. For $x, y \in P$, we denote their greatest lower bound by $x \wedge y$ and their least upper bound by $x \vee y$.

**Example 6.2.1**  In the poset of subsets of a set $S$ partially ordered by set inclusion, $x \wedge y$ is $x \cap y$ and $x \vee y$ is $x \cup y$.

**Example 6.2.2**  For the poset of the real numbers with the usual ordering, $x \wedge y$ is $\min(x, y)$ and $x \vee y$ is $\max(x, y)$.

**Example 6.2.3**  In the poset of the natural numbers partially ordered by divisibility, $x \wedge y$ is $\gcd(x, y)$, the greatest common divisor of $x$ and $y$ and $x \vee y$ is $\operatorname{lcm}(x, y)$, the least common multiple of $x$ and $y$.

Two posets $(P_1, \leq_1)$ and $(P_2, \leq_2)$ are said to be **isomorphic** if there is a bijective map $f : P_1 \to P_2$ such that $x \leq_1 y$ if and only if $f(x) \leq_2 f(y)$.

**Example 6.2.4**  Let $S$ be a set of $n$ elements and consider the poset $\mathcal{P}(S)$ of subsets of $S$. Let $I = \{0, 1\}$ be the two element poset defined by $0 < 1$. The correspondence between a subset and its characteristic vector is a bijective map that satisfies the previous property, and therefore

$$\mathcal{P}([n]) \simeq I^n, \forall n \in \mathbb{N}.$$

We compute now the Möbius function of $\mathcal{P}([n])$. Assume that $(P_1, \leq_1)$ and $(P_2, \leq_2)$ are two locally finite posets. It is not too difficult to check that the Möbius function of $P_1 \times P_2$ is given by

$$\mu((x_1, x_2), (y_1, y_2)) = \mu(x_1, y_1)\mu(x_2, y_2). \tag{6.2.1}$$

The Möbius function for $I$ is easily seen to be given by

$$\mu(x, y) = \begin{cases} (-1)^{y-x}, & \text{if } x \leq y \\ 0, & \text{otherwise.} \end{cases}$$

Thus, the Möbius function for $I^n$ is given by

$$\mu((x_1, .., x_n), (y_1, ..., y_n)) = (-1)^{\sum_{i=1}^{n}(y_i - x_i)},$$

whenever $x_i \leq y_i$ for any $1 \leq i \leq n$ and is 0, otherwise. Using the isomorphism between $\mathcal{P}([n])$ and $I^n$ given above, we deduce that

$$\mu(A, B) = \begin{cases} (-1)^{|B|-|A|}, & \text{if } A \subseteq B \\ 0, & \text{otherwise.} \end{cases}$$

The Möbius inversion formula for sets now reads as follows.

**Theorem 6.2.5**  *Let $S$ be a finite set. Let $f, g : \mathcal{P}(S) \to \mathbb{F}$. The following statements are equivalent:*

*(1)* $f(A) = \sum_{B:A \subseteq B} g(B), \quad \forall A \subseteq S.$
*(2)* $g(A) = \sum_{B:A \subseteq B} (-1)^{|B|-|A|} f(B), \quad \forall A \subseteq S.$

We can specialize this to deduce the principle of inclusion and exclusion. Indeed, suppose we have a set $A$ with *bad* subsets $A_i$ with $i \in [n]$. We would like to derive a formula for

$$|A \setminus \cup_{i=1}^{n} A_i|.$$

For each subset $J$ of $[n]$, we let $f(J)$ be the number of elements of $A$ which belong to every $A_j$, $j \in J$. Let $g(J)$ be those elements which belong to every $A_j$, $j \in J$ and in no other $A_i$ for $j \notin J$. Clearly,

$$f(K) = \sum_{J:J \supseteq K} g(J), \quad \forall K \subseteq [n].$$

By Möbius inversion, we obtain that

$$g(K) = \sum_{J:J \supseteq K} (-1)^{|J|-|K|} f(J), \quad \forall K \subseteq [n].$$

Note that $f(J) = |\cap_{j \in J} A_j|$. For $K = \emptyset$, we get that

$$g(\emptyset) = \sum_{J \subseteq [n]} (-1)^{|J|} f(J) = \sum_{J \subseteq [n]} |\cap_{j \in J} A_j|.$$

Since $g(\emptyset) = |A \setminus \cup_{i=1}^{n} A_i|$, this result proves the principle of inclusion and exclusion. Thus, the Möbius inversion formula is a vast generalization of this important principle.

**Example 6.2.6** Let us consider the lattice $D(n)$ of divisors of a natural number $n \geq 2$. By the unique factorization theorem, we see that if

$$n = p_1^{\alpha_1} \cdot \ldots \cdot p_k^{\alpha_k},$$

with $p_1, \ldots, p_k$ being distinct primes and $k \geq 1$, then

$$D(n) \simeq D(p_1^{\alpha_1}) \cdot \ldots \cdot D(p_k^{\alpha_k}).$$

Hence, to determine the Möbius function of $D(n)$, it suffices to determine it for $D(p^a)$ for a prime $p$ and a natural number $a$.

Let $p$ be a prime. Observe that $\mu(p^i, p^j) = 1$ if $i = j$, $-1$ if $i = j - 1$ and zero otherwise. By Eq. (6.2.1), the Möbius function for $D(n)$ is easily computed: $\mu(a, b) = 0$ unless $a|b$ in which case equals the classical Möbius function $\mu(b/a)$ defined as follows: $\mu(1) = 1$ and $\mu(n) = 0$ for $n \geq 2$ unless $n$ is the product of $k$ distinct primes in which case $\mu(n) = (-1)^k$.

The Möbius inversion formula for the lattice of natural numbers partially ordered by divisibility is now seen as an immediate consequence of the general inversion formula.

**Theorem 6.2.7** *Let $f, g : \mathbb{N} \to \mathbb{R}$ be two functions defined on the set of natural numbers. The following statements are equivalent:*

*(1)* $f(n) = \sum_{d:d|n} g(d), \forall n \in \mathbb{N}.$
*(2)* $g(n) = \sum_{d:d|n} \mu(d) f(n/d), \forall n \in \mathbb{N}.$

As an application of the Möbius inversion formula, one can obtain the formula for Euler's totient function $\phi(n)$. Recall that $\phi(n)$ denotes the number of natural numbers between 1 and $n$ that are coprime with $n$.

**Corollary 6.2.8** *For any natural number $n$,*

$$\phi(n) = \sum_{d|n} \mu(n/d)d.$$

***Proof*** The result follows from the observation that

$$\sum_{d:d|n} \phi(d) = n, \forall n \in \mathbb{N}$$

and the previous inversion formula.                                              ∎

**Example 6.2.9** There are many applications of the inversion formula in counting problems. For instance, let us look at the following celebrated example. If we have an infinite supply of beads of $\lambda$ colours, in how many ways can we make a necklace of $n$ beads? Clearly, any necklace can be thought of as a sequence $(a_1, \ldots, a_n)$, where we identify any cyclic permutation of the sequence as giving rise to the same necklace. We will say that a necklace is **primitive** of length $n$ if for no divisor $d < n$ it is not obtained by repeating $n/d$ times a necklace of length $d$. We say a necklace has **period** $d$ if it is obtained by repeating a primitive necklace of length $d$ for $n/d$ times. With these notions, we can count first the number of sequences to be $\lambda^n$. On the other hand, each sequence corresponds to some primitive necklace of period $d$, which must necessarily divide $n$. If we let $M(d)$ be the number of primitive necklaces of length $d$, we have $d$ places from which to start the sequence, and so we obtain

$$\lambda^n = \sum_{d:d|n} d M(d).$$

By Möbius inversion, we get that

$$M(n) = \frac{1}{n} \sum_{d:d|n} \mu(d) \lambda^{n/d}.$$

Now the total number of necklaces is

$$\sum_{d:d|n} M(d).$$

This can be simplified further. We have that

$$\sum_{d:de=n} \sum_{b:ab=d} \frac{\mu(a)\lambda^b}{d} = \sum_{b:b|n} \lambda^b \sum_{a:ae=n/b} \frac{\mu(a)}{a}.$$

The inner sum equals $\frac{n\phi(n/b)}{b}$. Thus, the final formula is

$$\frac{1}{n} \sum_{b:b|n} \phi(n/b)\lambda^b.$$

**Example 6.2.10**  Let $S$ be a finite set and $\Pi(S)$ the collection of its partitions. We make $\Pi(S)$ into a poset as follows. Recall that the components of a partition are called blocks. We say that $\alpha \leq \beta$ if every block of $\beta$ is contained in a block of $\alpha$. That is, each block of $\alpha$ is a union of blocks of $\beta$. For example,

$$\alpha = \{1, 2\}\{3, 4, 5\} \leq \{1\}\{2\}\{3, 5\}\{4\} = \beta.$$

It is easy to verify that this poset is a lattice with minimal element 0 given by the partition consisting of one block containing all the elements of $S$. The maximal element 1 is given by the partition consisting of singleton sets. Thus, the *greater* the partition, the *larger* the number of blocks.

We would like to determine the Möbius function of this lattice. To this end, let us define $b(\alpha)$ to be the number of blocks of the partition $\alpha$. Let us fix a partition $\beta$ with $m$ blocks. If $\alpha \leq \beta$, then every block of $\alpha$ is a union of blocks of $\beta$ and it is then clear that if we view $\beta$ as a set of its blocks, then

$$[0, \beta] \simeq \Pi(\beta),$$

which will be useful in the computation of the Möbius function.

Let $x$ be a variable/indeterminate. For each partition $\alpha$, define $g(\alpha)$ to be the polynomial $(x)_{b(\alpha)} = \prod_{j=0}^{b(\alpha)-1}(x - j)$. Using the results in Sect. 3.4, we have that

$$\sum_{\alpha \leq \beta} g(\alpha) = \sum_{\alpha \leq \beta}(x)_{b(\alpha)} = \sum_{k=1}^{m} S(m, k)(x)_k = x^m = x^{b(\beta)}.$$

By Möbius inversion and Sect. 3.4 again, we obtain that

$$g(\beta) = (x)_m = \sum_{\alpha \leq \beta} \mu(\alpha, \beta) x^{b(\alpha)} = \sum_{k=1}^{m} s(m, k) x^k.$$

Identifying the coefficients of $x^k$ of both sides of the identity above gives

$$s(m, k) = \sum_{\alpha \leq \beta, b(\alpha) = k} \mu(\alpha, \beta).$$

Taking $k = 1$ gives

$$s(m, 1) = \mu(0, \beta).$$

Thus, the value of the Möbius function $\mu(0, \beta)$ depends only on the number of blocks in $\beta$, namely, $b(\beta)$. But recall that $(-1)^{m-1} s(m, 1)$ is the number of permutations of $S_m$ with exactly one cycle in their disjoint cycle decomposition. The number of such permutations is $(m - 1)!$. Thus, we have proved the following result.

**Theorem 6.2.11** *For the lattice of partitions $\Pi(S)$ of a set $S$ with $n$ elements,*

$$\mu(0, 1) = (-1)^{n-1}(n - 1)!.$$

**Example 6.2.12** We will now count the number of connected labelled graphs on $n$ vertices. To this end, let us observe that any graph induces partition on the vertices given by its connected components. For each partition $\beta$ of the $n$ vertices, let $g(\beta)$ be the number of graphs whose partition of connected components is finer than $\beta$. Let $f(\beta)$ be the number of graphs whose partition of connected components is equal to $\beta$. Clearly,

$$g(\beta) = \sum_{\alpha \geq \beta} f(\alpha).$$

By Möbius inversion, we get that

$$f(\beta) = \sum_{\alpha \geq \beta} \mu(\beta, \alpha) g(\alpha).$$

What we want to determine is $f(0)$. But this is

$$f(0) = \sum_{\alpha} \mu(0, \alpha) g(\alpha).$$

If $\alpha = \{B_1, ..., B_k\}$, then clearly $g(\alpha) = 2^{\binom{|B_1|}{2}} \cdot .... \cdot 2^{\binom{|B_k|}{2}}$. Therefore

$$f(0) = \sum_{\alpha} (-1)^{b(\alpha)-1} (b(\alpha) - 1)! 2^{\binom{|B_1|}{2}} \cdot .... \cdot 2^{\binom{|B_k|}{2}}.$$

## 6.3  The Chromatic Polynomial

Graph colouring is one of the main topics in graph theory. In this section, we describe some connections between this subject and Möbius inversion.

Given a map $M$ in the plane and $\lambda$ a non-negative integer, denote by $p_M(\lambda)$ the number of colourings of $M$ using $\lambda$ colours such that no two neighbouring regions receive the same colour. Any such colouring is also called a **proper colouring** of $M$. We denote by $r(M)$ the number of regions of $M$. If we delete some (possibly none) borders of the map $M$, we obtain a *submap* of $M$. We can define a partial order on the collection of the *submaps* of $M$ as: $M_1 \leq M_2$ if $M_1$ is obtained from $M_2$ by deleting some of $M_2$'s borders. Clearly, $M_1 \leq M_2$ implies that $r(M_1) \geq r(M_2)$, but the converse may not be true. Our first result in this section is an application of Möbius inversion and implies that $p_M(\lambda)$ is a polynomial in $\lambda$ of degree $r(M)$.

**Theorem 6.3.1** *Let $M$ be a map. If $\mu$ be the Möbius function associated with poset of submaps of $M$, then*

$$p_M(\lambda) = \sum_{M':M' \leq M} \mu(M', M)\lambda^{r(M')},$$

*for any non-negative integer $\lambda$.*

**Proof** The total number of colourings of $M$ (proper and not proper colourings) equals $\lambda^{r(M)}$. Any such colouring of $M$ may be *refined* to a proper colouring of the unique *submap* of $M$ obtained by deleting the common boundary between any two regions that have the same colour. Hence,

$$\lambda^{r(M)} = \sum_{M':M' \leq M} p_{M'}(\lambda).$$

By applying Möbius inversion on the poset of submaps, we get the desired result. ∎

Given a graph $X = (V, E)$ that may contain multiple edges, but no loops, and a non-negative integer $\lambda$, denote by $p_X(\lambda)$ the number of proper colourings of $X$ using $\lambda$ colours. By a similar argument as before, we can derive the more general result that $p_X(\lambda)$ is a polynomial in $\lambda$ for any graph $X$. We give an alternative proof of this fact using induction and the following theorem. Recall that given a graph $X$ and an edge $e$, $X/e$ denotes the graph obtained from $X$ by contracting the endpoints $e$. Note that $X/e$ may have multiple edges.

**Theorem 6.3.2** *Let $X$ be a graph that may contain multiple edges, but no loops. If $\lambda$ is a non-negative integer and $e$ is an edge of $X$, then*

$$p_X(\lambda) = p_{X-e}(\lambda) - p_{X/e}(\lambda).$$

**Proof** Consider all proper colourings of $X - e$ with $\lambda$ colours. There are $p_{X-e}(\lambda)$ of them. These colourings can be split into two groups: the ones where the endpoints of

$e$ have different colours and the ones where the endpoints of $e$ have the same colour. The number of colourings in the first group is $p_X(\lambda)$ and the number of colourings in the second group is $p_{X/e}(\lambda)$. This proves result. ∎

We now prove that $p_X(\lambda)$ is a polynomial in $\lambda$.

**Theorem 6.3.3** *Let $X = (V, E)$ be a graph with $n$ vertices as in the previous theorem. For non-negative integer $\lambda$, $p_X(\lambda)$ is a monic polynomial in $\lambda$ of degree $n$ with integer coefficients.*

**Proof** We prove this result by strong induction on the number of edges and the previous theorem. The claim holds trivially if $X$ is the empty graph for then $p_X(\lambda) = \lambda^n$. For the induction step, let $X$ be a graph with at least one edge and assume that the statement is true for all graphs with fewer edges than $X$. By the induction hypothesis, we may write

$$p_{X-e}(\lambda) = \lambda^n + a_{n-1}\lambda^{n-1} + \cdots + a_0,$$

and

$$p_{X/e}(\lambda) = \lambda^{n-1} + b_{n-2}\lambda^{n-2} + \cdots + b_0,$$

for some integers $a_{n-1}, \ldots, a_0, b_{n-2}, \ldots, b_0$. Using the previous result, we get that

$$p_X(\lambda) = p_{X-e}(\lambda) - p_{X/e}(\lambda)$$

$$= \lambda^n + \sum_{j=0}^{n-1} a_j\lambda_j - \lambda^{n-1} - \sum_{\ell=0}^{n-2} b_\ell\lambda^\ell$$

$$= \lambda^n + (a_{n-1} - 1)\lambda^{n-1} + \sum_{k=0}^{n-2}(a_k - b_k)\lambda^k.$$

The theorem is proved. ∎

Based on this result, $p_X(\lambda)$ is called **the chromatic polynomial** of $X$. This polynomial was introduced by the American mathematician George David Birkhoff (1884–1944) in 1912 as an attempt to attack the Four Colour Conjecture (now Theorem). Showing that $p_X(4) > 0$ for any planar graph $X$ is equivalent to the Four Colour Theorem. While this approach did not lead to the proof of this theorem, the introduction of the chromatic polynomial opened up interesting mathematical research areas connecting combinatorics and statistical mechanics.

**Example 6.3.4** The chromatic polynomial of the complete graph $K_n$ is

$$\lambda(\lambda - 1)(\lambda - 2)\cdots(\lambda - (n-1)) = \sum_{k=0}^{n} s(n, k)\lambda^k,$$

where $s(n, k) = (-1)^{n-k}|s(n, k)|$ are the Stirling numbers of the first kind. Recall that $|s(n, k)|$ is the number of permutations of the symmetric group $S_n$ with exactly $k$-cycles in its unique factorization as a product of disjoint cycles.

Theorem 6.3.2 can be used to determine the chromatic polynomial of any tree.

**Theorem 6.3.5** *If $T$ be a tree with $n$ vertices, then*

$$p_T(\lambda) = \lambda(\lambda - 1)^{n-1}.$$

*Proof* We use induction on $n$. For $n = 1$, the statement is true. Let $T$ be a tree with $n \geq 2$ vertices. Assume that the statement is true for any tree with $n - 1$ or fewer vertices. The tree $T$ has a leaf $x$. Let $e$ be the unique edge containing vertex $x$. The graph $T - e$ consists of an isolated vertex and a tree with $n - 1$ vertices. Hence, its chromatic polynomial equals the product of the chromatic polynomials of these graphs, which is $\lambda \cdot \lambda(\lambda - 1)^{n-2} = \lambda^2(\lambda - 1)^{n-2}$, by the induction hypothesis. The graph $T/e$ is a tree with $n - 1$ vertices and $p_{T/e}(\lambda) = \lambda(\lambda - 1)^{n-2}$, again from the induction hypothesis. Hence,

$$\begin{aligned} p_T(\lambda) &= p_{T-e}(\lambda) - p_{T/e}(\lambda) \\ &= \lambda^2(\lambda - 1)^{n-2} - \lambda(\lambda - 1)^{n-2} \\ &= \lambda(\lambda - 1)^{n-1}. \end{aligned}$$

This proves the theorem.                                                        ■

It is rather remarkable that the converse of Theorem 6.3.5 also holds (see Exercise 6.6.13).

Theorems 6.3.2 and 6.3.5 can be used to determine the chromatic polynomial of the cycle $C_n$ on $n$ vertices. Deleting an edge from the cycle gives a tree on $n$ vertices and contracting an edge gives a cycle on $n - 1$ vertices. Thus, by an inductive argument, the motivated reader can deduce the following result.

**Theorem 6.3.6** *Let $n \geq 3$ be a natural number. The chromatic polynomial of the cycle $C_n$ is $(\lambda - 1)^n + (-1)^n(\lambda - 1)$.*

## 6.4  The Chromatic Number

**The chromatic number** $\chi(X)$ of the graph $X$ is the smallest non-negative integer $k$, so that $p_X(k) > 0$. In other words, $\chi(X)$ is the smallest number $k$ such that $X$ has a proper colouring with $k$ colours. Determining the chromatic number of a graph $X$ is in general a computationally difficult problem.

**Example 6.4.1** The chromatic number of the complete graph $K_n$ is $n$. The chromatic number of the empty graph $K_n^c$ is 1.

**Example 6.4.2** For the cycle graph $C_n$, the chromatic number is 2 or 3 accordingly as $n$ is even or odd.

To get an upper bound $\chi(X) \leq s$, one must construct or show the existence of a proper colouring of $X$ with $s$ colours. The Four Colour Theorem is the assertion that the chromatic number of any graph obtained from a planar map is at most 4. The following result gives a general upper bound for the chromatic number of a graph.

**Theorem 6.4.3** *If $X = (V, E)$ is a graph with maximum degree $\Delta$, then $\chi(X) \leq 1 + \Delta$.*

**Proof** We list the vertices of $X$ in some order $v_1, \ldots, v_n$. We construct a *greedy* colouring of $X$ with colours $\{1, \ldots, \Delta + 1\}$ by colouring the vertices in the order above, assigning to $v_i$ the smallest indexed colour not already used by its neighbours $v_j$ with $j < i$. Each vertex $v_i$ will have at most $d(v_i) \leq \Delta$ such neighbours $v_j$ with $j < i$, so this colouring will not use more than $\Delta + 1$ colours. ∎

This theorem is sharp for complete graphs and odd cycles. These are the only graphs where equality is attained, as shown by the English mathematician Rowland Leonard Brooks (1916–1993) in 1941.

**Theorem 6.4.4** (Brooks, 1941) *If $X$ is a graph with maximum degree $\Delta$ that is not a complete graph nor an odd cycle, then $\chi(X) \leq \Delta$.*

We will now connect eigenvalues of the adjacency matrix of a graph with its chromatic number. As preparation to this end, we will review the notion of **Rayleigh-Ritz quotient** from linear algebra. For a real $n \times n$ matrix $A$ and a non-zero vector $v \in \mathbb{R}^n$, we call $\frac{v^T A v}{v^T v}$ the **Rayleigh-Ritz quotient of** $v$.

**Theorem 6.4.5** *Let $A$ be a real and symmetric $n \times n$ matrix. If $\lambda_{\max}$ and $\lambda_{\min}$ are the largest and smallest eigenvalues of $A$, respectively, then*

$$\lambda_{\max} = \max_{v \in \mathbb{R}^n, v \neq 0} \frac{v^T A v}{v^T v}, \tag{6.4.1}$$

*and*

$$\lambda_{\min} = \min_{v \in \mathbb{R}^n, v \neq 0} \frac{v^T A v}{v^T v}. \tag{6.4.2}$$

**Proof** Let $\lambda_1 \geq \ldots \geq \lambda_n$ be the eigenvalues of $A$. If $U$ is a matrix whose columns form an orthonormal basis of eigenvectors of $A$, then we may write $A = U D U^T$, where $D$ is a diagonal matrix whose diagonal entries are the eigenvalues of $A$. Let $v$ be a non-zero vector in $\mathbb{R}^n$. We have that

$$v^T A v = v^t U D U^t v = \sum_{i=1}^{n} \lambda_i |(U^T v)_i|^2$$

$$\leq \lambda_{max} \sum_{i=1}^{n} |(U^T v)_i|^2.$$

Since $U$ is an orthogonal matrix, we have that

$$\sum_{i=1}^{n} |(U^T v)_i|^2 = \sum_{i=1}^{n} |v_i|^2 = v^T v.$$

Thus, $\lambda_{max} \geq \frac{v^T A v}{v^T v}$. If $v$ is an eigenvector corresponding to $\lambda_{max}$, we get equality. This proves the assertion related to $\lambda_{max}$. We leave the proof of the $\lambda_{min}$ as an exercise. ∎

If $X$ is a graph, let us denote $\lambda_{max}(X)$ and $\lambda_{min}(X)$ to be the largest and smallest eigenvalues of the adjacency matrix $A$ of $X$. Recall that $Y$ is a **subgraph of** $X$ if $Y$ is obtained from $X$ by deleting some vertices and edges.

**Corollary 6.4.6** *If $Y$ is a subgraph of a graph $X$, then*

$$\lambda_{max}(Y) \leq \lambda_{max}(X); \quad \lambda_{min}(Y) \geq \lambda_{min}(X).$$

*Proof* The first part of the theorem is proved as follows. By relabelling the vertices, we may assume that the adjacency matrix $A$ of $X$ has a leading principal submatrix $B$ which is the adjacency matrix of $Y$. Let $v$ be a non-zero vector of the appropriate dimension such that $Bv = \lambda_{max}(Y)v$. Let $u$ be the column vector with $|V(X)|$ rows formed by adjoining zero to entries of $v$. Then,

$$\lambda_{max}(Y) = \frac{v^T B v}{v^T v} = \frac{u^T A u}{u^T u} \leq \lambda_{max}(A).$$

This proves the first assertion. The other inequality is proved similarly, and we leave the details as an exercise. ∎

Another consequence of Theorem 6.4.5 is the following inequality involving the largest eigenvalue of the adjacency matrix and the average degree of a graph (see also Exercise 4.6.11).

**Corollary 6.4.7** *Let $X$ be a graph average degree $\overline{d}$ and maximum degree $\Delta$. If $\lambda_{max}$ is the largest eigenvalue of its adjacency matrix, then*

$$\overline{d} \leq \lambda_1 \leq \Delta. \tag{6.4.3}$$

An immediate consequence is that $\lambda_1 \geq \delta(X)$, where $\delta(X)$ denotes the minimum degree of $X$. We say a graph $X$ is $t$-**critical** if $\chi(X) = t$ and for any proper subgraph $Y$ of $X$, $\chi(Y) < t$.

**Lemma 6.4.8** *If $X$ be a graph with chromatic number $t \geq 2$, then $X$ has a $t$-critical subgraph $Y$ with minimum degree $\delta(Y) \geq t - 1$.*

*Proof* The set of all subgraphs of $X$ is non-empty and contains some graphs (for instance, $X$ itself) that have chromatic number $t$. Let $Y$ be a subgraph of $X$ with chromatic number $t$ such that $Y$ is minimal with respect to the number of vertices. Clearly, $Y$ is $t$-critical. Moreover, if $y \in V(Y)$, then the graph $Y - y$ is a subgraph of $Y$ and has a vertex colouring with $t - 1$ colours. If $y$ had $t - 2$ or fewer neighbours in $Y$, then we could have extended the proper colouring with $t - 2$ colours of $Y - y$ to a proper colouring of $Y$ with $t - 1$ colours. This gives a contradiction with $\chi(Y) = t$. Hence, $\delta(Y) \geq t - 1$ and this finishes the proof. ∎

The previous lemma has the following important consequence that was proved by the Hungarian-Australian mathematician George Szekeres (1911–2005) and the American mathematician Herbert Wilf (1931–2012) in 1968 (see Szekeres and Wilf (1968)).

**Theorem 6.4.9** (Szekeres-Wilf 1968) *If $X$ is a graph, then*

$$\chi(X) \leq 1 + \max_{Y \subseteq X} \delta(Y),$$

*where the maximum is taken over all subgraphs $Y$ of $X$.*

**Proof** By Lemma 6.4.8, there is a vertex subgraph $Z$ of $X$ whose chromatic number is $\chi(Z)$ and $\delta(Z) \geq \chi(X) - 1$. Thus, we have

$$\chi(X) \leq 1 + \delta(U) \leq 1 + \max_{Y \subseteq X} \delta(Y).$$

This proves the theorem. ∎

By a slight modification of the previous proof, we also get the following result.

**Theorem 6.4.10** (Wilf 1967) *For any graph $X$, we have*

$$\chi(X) \leq 1 + \lambda_{\max}(X).$$

**Proof** As before, there is a vertex subgraph $Z$ of $X$ whose chromatic number is $\chi(X)$ and $\delta(Z) \geq \chi(X) - 1$. Thus, by Theorem 6.4.6, we have that

$$\chi(X) \leq 1 + \delta(Z) \leq 1 + \lambda_{\max}(Z) \leq 1 + \lambda_{\max}(X),$$

as desired. ∎

Let $X$ be a graph and $r$ be a natural number. Proving that $\chi(X) > r$ is equivalent to showing that no colouring of $X$ with $r$ colours is proper. Such strategy can be used, for example, to show that $\chi(C_{2k+1}) > 2$ for any $k \geq 1$. There are other general bounds that one can also use.

**Proposition 6.4.11** *If $X$ is a graph whose clique number is $\omega(X)$, then $\chi(X) \geq \omega(X)$.*

**Proof** Let $Y$ be a complete subgraph of $X$ with $\omega(X)$ vertices. Any proper colouring of $X$ must use different colours on the vertices of $Y$. Hence, $\chi(X) \geq \omega(X)$. ∎

**Fig. 6.4** An empty Sudoku
puzzle

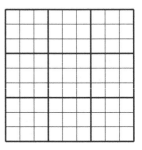

For some graphs (complete graphs, bipartite graphs, or Sudoku graphs), this bound
can give the correct value of $\chi(X)$, but there are graphs with no triangles having
arbitrarily large chromatic number.

**Proposition 6.4.12** *If $X = (V, E)$ is a graph with $n$ vertices and independence
number $\alpha$, then $\chi(X) \geq n/\alpha$.*

**Proof** Let $\chi(X) = t$ and consider a partition of $V$ into $t$ independent sets $X_1, \ldots, X_t$.
It follows that

$$n = |V| = \sum_{j=1}^{t} |X_j| \leq t\alpha,$$

which is the desired result.                                                              ∎

## 6.5  Sudoku Puzzles

The Sudoku puzzle has become a popular puzzle that many newspapers carry as a
daily feature. The puzzle consists of a $9 \times 9$ square grid in which some entries of the
grid have a number from 1 to 9. One is then required to complete the grid in such a
way that every row, every column, and every one of the nine $3 \times 3$ *sub-grids* contain
each digit from 1 to 9. The *sub-grids* are shown below (Fig. 6.4).

For anyone trying to solve a Sudoku puzzle, several questions arise naturally.
For a given puzzle, does a solution exist? If the solution exists, is it unique? If it is
not unique, how many solutions are there? Moreover, is there a systematic way of
determining all the solutions? What is the minimum number of entries that can be
specified in a puzzle to ensure a unique solution. We leave it to the reader to show that
the puzzle below has a unique solution. Thus, the minimum number is at most 17. In
2012, Gary McGuire, Bastian Tugemann, and Gilles Civario completed a computer
search for 16-clue Sudoku puzzles[1] without finding any such puzzles, thus showing
that the answer is 17 (Fig. 6.5).

---

[1] Their paper *There is no 16-Clue Sudoku: Solving the Sudoku Minimum Number of Clues Problem*
is available at https://arxiv.org/abs/1201.0749.

**Fig. 6.5** A partial Sudoku puzzle with 17 entries

| | | | | | | | | 1 |
|---|---|---|---|---|---|---|---|---|
| 4 | | | | | | | | |
| | 2 | | | | | | | |
| | | | | 5 | | 4 | | 7 |
| | | | 8 | | | 3 | | |
| | | | 1 | | 9 | | | |
| 3 | | | | 4 | | | 2 | |
| | 5 | | 1 | | | | | |
| | | | | 8 | | 6 | | |

**Fig. 6.6** A partial Sudoku puzzle with two solutions

| 9 | 6 | | 7 | | | 4 | | 3 |
|---|---|---|---|---|---|---|---|---|
| | | | 4 | | | 2 | | |
| | 7 | | | 2 | 3 | | 1 | |
| 5 | | | 2 | | | 1 | | |
| | 4 | | | | 8 | | 6 | |
| | | 3 | | | | | | 5 |
| | 3 | | 7 | | | | 5 | |
| | 7 | | | | 5 | | | |
| 4 | 5 | | 1 | | | 7 | | 8 |

We can interpret the Sudoku puzzle as a vertex colouring problem. We associate a graph with the $9 \times 9$ Sudoku grid as follows. The vertices correspond to the 81 cells of the grid. Two vertices/cells are adjacent if and only if the corresponding cells in the grid are in the same row or in the same column or in the same sub-grid. Each complete Sudoku grid corresponds to a proper colouring of this graph with colours $\{1, \ldots, 9\}$.

One can generalize this problem to a $n^2 \times n^2$ grid which is partitioned into $n^2$ sub-grids of dimensions $n \times n$. Construct a graph $X_n$ associated to this grid as follows. The vertices are the ordered pairs $(i, j)$ with $0 \le i, j \le n^2 - 1$. Two vertices $(i, j)$ and $(i', j')$ are adjacent if $i = i'$ or $j = j'$ or $\lfloor i/n \rfloor = \lfloor i'/n \rfloor$ and $\lfloor j/n \rfloor = \lfloor j'/n \rfloor$. We call $X_n$ **the Sudoku graph of dimension** $n$. One can show that $X_n$ is regular of degree $3n^2 - 2n - 1$. When $n = 3$, the graph $X_3$ has 81 vertices and is 20-regular. When $n = 2$, the graph $X_2$ has 16 vertices and is 7-regular.

**Theorem 6.5.1** *For any natural number $n$, $\chi(X_n) = n^2$.*

**Proof** Any $n^2$ vertices/cells in the same row form a clique in $X_n$. Hence, $\omega(X_n) \ge n^2$. Actually equality holds, and we leave this as an exercise. By Proposition 6.4.11, $\chi(X_n) \ge \omega(X_n) = n^2$. For $0 \le i \le n$, write $i = nq_i + r_i$, where $0 \le q_i \le n - 1$ and $0 \le r_i \le n - 1$. Colour the vertex/cell $(i, j)$ by the colour $r_i n + q_i + nq_j + r_j$ (mod $n^2$). The reader can check that this is a proper colouring of $X_n$ which implies that $\chi(X_n) \le n^2$ and finishes our proof. ∎

Given a partial proper colouring $c$ of a given graph $X$, one can show that the number of ways of completing this colouring to obtain a proper colouring of $X$ with $\lambda$ colours is a polynomial in $\lambda$, provided that $\lambda$ is greater than the number of colours used in the partial colouring $c$. A Sudoku puzzle corresponds to a partial colouring of $X_n$ and the question is whether this partial colouring can be completed to a proper colouring of the Sudoku graph $X_n$ with $n^2$ colours.

It is not obvious at a first glance if a given puzzle has a solution. Also, it is not always clear whether a puzzle has a unique solution. An obvious necessary condition to have a unique solution is that the partial Sudoku square must contain at least 8 distinct numbers from $\{1, \ldots, 9\}$. This is not sufficient, as the square in Fig. 6.6 has exactly two solutions.

## 6.6 Exercises

**Exercise 6.6.1** Show that the examples from Sect. 6.1 are actually posets.

**Exercise 6.6.2** Draw the Hasse diagram for $S_4$ with the Bruhat order and determine completely the Möbius function of this poset.

**Exercise 6.6.3** Let $F, G : [1, +\infty) \to \mathbb{R}$ be two functions. If

$$G(x) = \sum_{n \leq x} F(x/n), \forall x \geq 1,$$

then prove that

$$F(x) = \sum_{n \leq x} \mu(n) G(x/n), \forall x \geq 1.$$

**Exercise 6.6.4** Let $x \geq 1$ be a real number. Show that

$$\sum_{n \leq x} \mu(n) \lfloor x/n \rfloor = 1.$$

**Exercise 6.6.5** Let $(P_1, \leq_1)$ and $(P_2, \leq_2)$ be two locally finite posets. Show that

$$\mu((x_1, y_1), (x_2, y_2)) = \mu(x_1, y_1) \mu(x_2, y_2).$$

**Exercise 6.6.6** Let $(P, \leq)$ be a finite poset. For $a \in P$, we will denote by $\downarrow a$ the set $\{x \in P : x \leq a\}$ and $\uparrow a$ the set $\{x \in P : a \leq x\}$. We say that $P$ is **linearly ordered** if any two elements of $P$ are comparable. Show that any partial ordering of $P$ can be extended to a linear ordering, as follows. View the poset $(P, \leq)$ as a subset $R$ of $P \times P$ satisfying the axioms: (1) $(a, a) \in R$, (2) $(a, b) \in R$ and $(b, a) \in R$ implies $a = b$ and (3) $(a, b) \in R$, $(b, c) \in R$ implies $(a, c) \in R$. A linear order can be regarded as a subset $R'$ of $P \times P$ which has the additional property that for any $a, b \in P$ either $(a, b) \in R'$ or $(b, a) \in R'$. Let now $a, b$ be incomparable in $(P, \leq)$. Put $R' = R \cup (\downarrow a \times \uparrow b)$. Verify that $R'$ is a partial order of $P$ in which $(a, b) \in R'$. Deduce that any partial ordering of $P$ can be extended to a linear ordering.

**Fig. 6.7** A graph on six
vertices

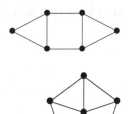

**Fig. 6.8** The wheel graph
$W_6$

**Exercise 6.6.7** Let $X$ be a graph and denoted by $p_X(\lambda)$ its chromatic polynomial.

(1) If $X_1, \ldots, X_t$ are the components of $X$, then

$$p_X(\lambda) = \prod_{i=1}^{t} p_{X_i}(\lambda).$$

(2) Prove that the constant term of $p_X(\lambda)$ is zero.
(3) Prove that the sum of the coefficients of the chromatic polynomial of $X$ is zero unless $X$ has no edges.
(4) Show that the coefficients of the chromatic polynomial of $X$ alternate in sign.

**Exercise 6.6.8** Compute the chromatic polynomial of the graph in Fig. 6.7.

**Exercise 6.6.9** Let $X$ and $Y$ be two graphs whose intersection is a complete graph $K_r$. If $Z$ is the graph whose edge set is the union of the edges of $X$ and $Y$, prove that

$$p_Z(\lambda) = \frac{p_X(\lambda) \cdot p_Y(\lambda)}{\lambda(\lambda - 1) \ldots (\lambda - r + 1)}.$$

**Exercise 6.6.10** For $n \geq 4$, **the wheel graph** $W_n$ is the graph obtained from the cycle graph on $n - 1$ vertices by adding a new vertex adjacent to each vertex of the cycle. Determine the chromatic polynomial of the wheel graph $W_n$ (Fig. 6.8).

**Exercise 6.6.11** Let $X$ be a connected graph with $n \geq 3$ vertices. Prove that $|p_X(\lambda)| \leq \lambda(\lambda - 1)^{n-1}$ for any non-negative integer $\lambda$.

**Exercise 6.6.12** Let $X$ be a graph with $n$ vertices. Prove that if $p_X(\lambda) = \lambda(\lambda - 1)^{n-1}$, then $X$ is a tree.

**Exercise 6.6.13** Let $n$ be a natural graph. Show that

$$p_{K_{2,n}}(\lambda) = \lambda(\lambda - 1)(\lambda - 2)^n + \lambda(\lambda - 1)^n.$$

**Fig. 6.9** The Petersen graph

**Exercise 6.6.14** Let $c(X)$ denote the number of components of the graph $X$ and for $F \subseteq E(X)$, denote by $X[F]$ the spanning subgraph of $X$ with edge set $F$. Show that

$$p_X(\lambda) = \sum_{F \subseteq E(X)} (-1)^{|F|} \lambda^{c(X[F])}.$$

**Exercise 6.6.15** Six different television stations are applying for channel frequencies, and no two stations can use the same frequency if they are within 150 miles of each other. If the distances between the stations $A, B, C, D, E$, and $F$ are given by the matrix below, find the minimal number of frequencies needed.

|   | A | B | C | D | E | F |
|---|---|---|---|---|---|---|
| A | − | 85 | 175 | 200 | 50 | 100 |
| B | 85 | − | 125 | 175 | 100 | 160 |
| C | 175 | 125 | − | 100 | 200 | 250 |
| D | 200 | 175 | 100 | − | 210 | 220 |
| E | 50 | 100 | 200 | 210 | − | 100 |
| F | 100 | 160 | 250 | 220 | 100 | − |

**Exercise 6.6.16** The **join** of two graphs $X$ and $Y$ is defined as the graph obtained by joining every vertex of $X$ to every vertex of $Y$. We denote this graph by $X \vee Y$. Show that $\chi(X \vee Y) = \chi(X) + \chi(Y)$.

**Exercise 6.6.17** Let $X$ be a graph with $n$ vertices. The **complement** $\overline{X}$ has the same vertex set as $X$, and its edge set is the complement of the edge set of $X$. Prove that $\chi(X) \cdot \chi(\overline{X}) \geq n$ and that $\chi(X) + \chi(\overline{X}) \leq n + 1$.

**Exercise 6.6.18** Let $X$ be a graph with $m$ edges. Show that $\chi(X) \leq \frac{-1+\sqrt{8m+1}}{2}$.

**Exercise 6.6.19** Let $X_n$ be the graph with vertex set $\{1, 2, \ldots, 2n\}$ with the adjacency relation given by $(i, j)$ is an edge if and only if $i$ and $j$ have a common prime divisor. Show that the chromatic number of $X_n$ is at least $n$ (Fig. 6.9).

**Exercise 6.6.20** The **Kneser graph** $K(n, k)$ is the graph whose vertices are all the $k$-element subsets of $[n]$. Two $k$-subsets are adjacent in $K(n, k)$ if and only if they are disjoint. Show that the Petersen graph is isomorphic to $K(5, 2)$ and that $\chi(K(n, k)) \leq n - 2k + 2$.

# Reference

G. Szekeres, H. Wilf, An inequality for the chromatic number of a graph. J. Comb. Theory **4**, 1–3 (1968)

# Chapter 7
# Enumeration Under Group Action

## 7.1 Basic Facts About Groups

**A group** $(G, *)$ consists of a set $G$ together with a binary operation $*$ on $G$ satisfying the following axioms:

(1) (closure): For any $a, b \in G$, $a * b \in G$.

(2) (associativity): For any $a, b, c \in G$

$$(a * b) * c = a * (b * c).$$

(3) (identity element): There is an element called the **identity element** $e \in G$ such that
$$a * e = e * a = a, \forall a \in G.$$

(4) (inverse element): For any $a \in G$, there is a $b \in G$ so that

$$a * b = b * a = e.$$

We write $a^{-1}$ to denote the inverse of $a$.

If in addition to this, $a * b = b * a$ for all $a, b \in G$, we say that $G$ is **Abelian** or **commutative**. The name *Abelian* is used in honour of the Norwegian mathematician Niels Abel (1802–1829). When $G$ is finite, **the order** of $G$ is defined as the number of elements of $G$.

Note that in a group, we have the **cancellation law**:

$$a * b = a * c \text{ implies } b = c,$$

since we can multiply both sides on the left by $a^{-1}$. The cancellation law implies that the identity element is unique because if there were two identity elements $e$ and

© Hindustan Book Agency 2022

S. M. Cioabă and M. R. Murty, *A First Course in Graph Theory and Combinatorics*, Texts and Readings in Mathematics 55, https://doi.org/10.1007/978-981-19-0957-3_7

$e'$ say, then $a = a * e = a * e'$ which means that $e = e'$. Note that if $a * b = c * a$, we cannot necessarily conclude that $b = c$ (see Example 7.1.4).

The reason for studying groups in the abstract is that many scientific discoveries can be formulated in the language of group theory. In addition, the fundamental particles in the heart of the atom seem to know everything about non-Abelian groups! In fact, the character theory of certain subgroups of the group $GL_2(\mathbb{C})$ (see Example 7.1.4) led to the discovery of new sub-atomic particles in the early twentieth century.

We give now some examples of groups.

**Example 7.1.1** The group $(\mathbb{Z}, +)$ consists of the set of integers with the binary operation of addition. It is an Abelian group with identity element 0. Each of the set of rational numbers $\mathbb{Q}$ or real numbers $\mathbb{R}$ or complex numbers $\mathbb{C}$ forms a group with the corresponding operation of addition.

**Example 7.1.2** The set of non-zero integers $\mathbb{Z}^*$ with the binary operation of multiplication is not a group. This is because not every integer has a (multiplicative) inverse with this operation.

**Example 7.1.3** The group $(\mathbb{Q}^*, \cdot)$ consists of the set of rational numbers with the binary operation of multiplication. This is an Abelian group with identity element 1. Each of the set of non-zero reals $\mathbb{R}^*$ or the set of complex numbers $\mathbb{C}^*$ forms a group with the corresponding operation of multiplication.

All the previous examples have been infinite Abelian groups. We give now some examples of infinite non-Abelian groups.

**Example 7.1.4** The set $GL_2(\mathbb{R})$ consists of all $2 \times 2$ invertible matrices with entries in $\mathbb{R}$. This is a group with the binary operation of matrix multiplication. Similarly, the set $GL_2(\mathbb{C})$ of $2 \times 2$ invertible complex matrices forms a group with matrix multiplication. These are infinite non-Abelian groups. Notice that

$$\begin{bmatrix} 0 & -1 \\ 1 & 0 \end{bmatrix} \begin{bmatrix} a & b \\ c & d \end{bmatrix} = \begin{bmatrix} -c & -d \\ a & b \end{bmatrix} = \begin{bmatrix} d & -c \\ -b & a \end{bmatrix} \begin{bmatrix} 0 & -1 \\ 1 & 0 \end{bmatrix},$$

and we **cannot** cancel the matrix

$$\begin{bmatrix} 0 & -1 \\ 1 & 0 \end{bmatrix}$$

from both sides of the equation!

We give now some examples of finite groups.

**Example 7.1.5** Let $n \geq 2$ be a natural number. The set of integers modulo $n$, denoted by $\mathbb{Z}_n$ or $\mathbb{Z}/n\mathbb{Z}$, forms an Abelian group with addition modulo $n$.

**Example 7.1.6** The set of non-zero integers $\mathbb{Z}_6^*$ modulo 6 is not a group with the operation of multiplication modulo 6 since the residue class 2 does not have an inverse.

**Example 7.1.7** Let $n \geq 2$ be a natural number. The set of non-zero residue classes $\mathbb{Z}_n^*$ modulo $n$ forms a group with the multiplication modulo $n$ if and only if $n$ is a prime. In such case, the group is Abelian of order $n - 1$.

**Example 7.1.8** Let $n \geq 2$ be a natural number. The set of coprime residue classes modulo $n$ forms a group under multiplication. To show this, one needs to use the Euclidean algorithm (exercise). We sometimes denote this group as $(\mathbb{Z}/n\mathbb{Z})^\times$ or $\mathbb{Z}_n^\times$.

**Example 7.1.9** Let $n \geq 2$. The set $S_n$ of permutations of $[n]$ forms a group with operation of function composition. This is called **the symmetric group of order** $n$ and is not Abelian when $n \geq 3$.

To indicate $a * b$ we sometimes drop the $*$ and simply write $ab$ when there is no cause for confusion. There is a general tendency to use the multiplicative notation for writing the group binary operation, although there is no universal convention about this. Part of the reason for using multiplicative notation is to emphasize that the groups we are dealing with need not be Abelian. There is also a tendency to use the symbol 1 to denote the identity element (and 0 when we write the group additively).

**Example 7.1.10** Let $n \geq 3$ be a natural number. Consider a regular $n$-gon $P$. The rotation by $\frac{2\pi}{n}$ clockwise denoted $r$ and the reflection or flip about an axis of symmetry (say the vertical axis) $f$ generate a non-Abelian group of order $2n$. This is called **the dihedral group of order** $2n$ and is denoted by $D_{2n}$. When $n = 3$, $D_6$ is isomorphic to the group $S_3$ of permutations of $[3]$.

Given a group $(G, *)$, a subset $H$ of $G$ is called **a subgroup of** $G$, which is denoted by $H \leq G$, if $(H, *)$ is a group. **The order** $\mathrm{ord}(g)$ of an element $g \in G$ is the smallest natural number $n$ such that $g^n = 1$ if such natural numbers exist or $\infty$, otherwise. A group $(G, *)$ is called **cyclic** if there is an element $g \in G$ such that every element of the group is of the form $g^m$, for some integer $m$. For instance, $(\mathbb{Z}, +)$ is a cyclic group under addition with generator 1. As any cyclic group is finite or countable, $(\mathbb{R}, +)$ and $(\mathbb{R}^*, \cdot)$ are not cyclic groups. The additive group $(\mathbb{Z}_n, +)$ of residue classes mod $n$ is a cyclic group with generator being the residue class 1. Any coprime residue class will also serve as a generator.

**Theorem 7.1.11** *Let $G$ be a finite group.*

*(1) If $H$ is a subgroup of $G$, then $|H|$ divides $|G|$.*
*(2) If $g \in G$, then $\mathrm{ord}(g)$ divides $|G|$.*
*(3) If $|G| = n$, then $g^n = 1$ for any $g \in G$.*

***Proof*** For (1), define a binary relation $\sim$ on the set $G$, where $x \sim y$ if and only if $y \in xH$. This is an equivalence relation as it satisfies the conditions of being reflexive, symmetric, and transitive. As such, the set $G$ is partitioned into equivalence classes $y_1 H, \ldots, y_t H$ for some $t \geq 1$ and $y_1, \ldots, y_t \in G$. These equivalence classes are called **left cosets**. Since $|H| = |y_j H|$ for any $1 \leq j \leq t$, we get that

$$|G| = \sum_{j=1}^{t} |y_j H| = t|H|,$$

which proves our assertion. For (2), if $H$ is the subgroup of $G$ generated by $g$, then $|H| = \mathrm{ord}(g)$. By part (1), we get the desired result. Part (3) follows from $g^{\mathrm{ord}(g)} = 1$ and $\mathrm{ord}(g)|n$.                                                                          ∎

Part (1) of the previous theorem is named after the French-Italian mathematician Joseph Louis Lagrange (or Giuseppe Luigi Lagrangia) (1736–1813) and is usually called Lagrange's theorem. Part (3) of the previous result also implies Fermat's little theorem stating that if $p$ is prime and $a$ is coprime to $p$, then $a^{p-1} \equiv 1(\mod p)$.

## 7.2  The Orbit-Stabilizer Formula

Let $G$ be a group and $X$ a set. We say that **the group $G$ acts on $X$** if there is a map $G \times X \to X$ (usually denoted by $(g, x) \mapsto g \cdot x$) satisfying the following axioms for all $x \in X$:

(1)  $1 \cdot x = x$, where 1 is the identity element of $G$,
(2)  $(gh) \cdot x = g \cdot (h \cdot x)$ for all $g, h \in G$.

Here are a few examples of group actions.

**Example 7.2.1**  If $G$ is a group and $H$ is a subgroup, let $X = \{aH : a \in G\}$ be the set of left cosets of $H$ in $G$. The group $G$ acts on $X$ via $g(aH) = (ga)H$ for $g, a \in G$.

**Example 7.2.2**  If $G$ is a group, and we let $X$ be $G$ itself, then $G$ acts on itself via conjugation: $g \cdot x = gxg^{-1}$ for $g, x \in G$.

**Example 7.2.3**  Let $p$ be prime and $G = \mathbb{Z}_p$ be the additive group of residue classes mod $p$. Let $X$ be the set of all ordered $p$-tuples $(x_1, x_2, \ldots, x_p)$ where $x_i \in \{1, 2, \ldots, n\}$. Since $G$ is cyclic, it suffices to define how 1 acts on $X$. We put

$$1 \cdot (x_1, x_2, \ldots, x_p) = (x_p, x_1, \ldots, x_{p-1}).$$

In other words, 1 acts like a shift operator, shifting the coordinates by one component.

**Example 7.2.4**  Let $n$ be a natural number and $G = \mathbb{Z}_n$. Let $X$ be the set of all $n$-tuples $(x_1, \ldots, x_n)$ where $x_i \in \{1, 2, \ldots, \lambda\}$. We define

$$1 \cdot (x_1, x_2, \ldots, x_n) = (x_n, x_1, \ldots, x_{n-1}).$$

We can view the set $X$ as all the possible *necklaces* formed by using beads of $\lambda$ colours. This perspective will be useful in later applications.

It will be convenient to simplify our notation slightly. Instead of writing $g \cdot x$, we will simply write $gx$, when it is clear that $g \in G$ and $x \in X$. An action of $G$ on $X$ determines an equivalence relation on $X$ as follows. Namely, we will write $x \sim y$ if there is an element $g \in G$ such that $gx = y$. Thus, if $gx = y$ then $x = g^{-1}y$ and $y \sim x$. Since $1x = x$, this means that $x \sim x$. Also, it is easy to check that $x \sim y$ and $y \sim z$ implies $x \sim z$. Therefore, $\sim$ defines an equivalence relation on $X$. Consequently, we can partition $X$ into equivalence classes, which we call **orbits**. The orbit containing the element $x$ is

$$Gx := \{gx : g \in G\}.$$

If $G$ and $X$ are finite, it is natural to ask how many elements are in each orbit and how many orbits are there. We begin with the first question. We begin by figuring out when we get any repeats in the list:

$$gx, g \in G. \tag{7.2.1}$$

Note that $gx = hx$ if and only if $g^{-1}hx = x$, that is, if and only if $g^{-1}h$ fixes $x$. This leads to the notion of the **stabilizer** of $x$, denoted $G_x$, and defined as

$$G_x := \{g \in G : gx = x\},$$

the set of elements of $G$ fixing $x$. For any $x \in X$, the stabilizer $G_x$ is a subgroup of $G$. In the context above, we see that $gx = hx$ if and only if $h^{-1}g \in G_x$ which is the same as $gG_x = hG_x$. Thus, in the list (7.2.1), each element is repeated the same number of times, namely, $|G_x|$ times so that the number of distinct elements is $|G|/|G_x|$.

**Theorem 7.2.5** (Orbit-Stabilizer Formula) *Let $G$ be a finite group acting on a finite set $X$. For any $x \in X$,*
$$|Gx| = |G|/|G_x|. \tag{7.2.2}$$

It is clear that $X$ will be partitioned into the orbits $Gx_1, \ldots, Gx_t$, for some $t \geq 1$ and $x_1, \ldots, x_t \in X$. Therefore,

$$|X| = \sum_{i=1}^{t} |Gx_i|.$$

For each subgroup $H$ of $G$, we define fix$(H)$ to be the set of $H$-fixed points of $X$:

$$\mathrm{fix}(H) = \{x \in X : \quad hx = x \quad \forall h \in H\}.$$

If $g \in G$, we simply write fix$(g)$ for the set of elements fixed by the subgroup generated by $g$. From the above relation, we separate those $x_i$'s for which $Gx_i$ consists of singleton sets and obtain the following theorem.

**Theorem 7.2.6** *If $G$ is a finite group acting on a finite set $X$, then*

$$|X| = |\text{fix}(G)| + \sum_{i:G_{x_i} \neq G} |G|/|G_{x_i}|.$$

This formula has many applications. Given a group $G$, let $Z(G) = \{g : gx = xg, \forall x \in G\}$ denote its **centre**. For $x \in G$, denote by $C(x) = \{g \in G : gx = xg\}$ **the centralizer** of $x$ in $G$. When a group $G$ acts on itself via conjugation, the Orbit-Stabilizer Formula has the following consequence.

**Corollary 7.2.7** (The Class Equation) *If $G$ is a finite group, then*

$$|G| = |Z(G)| + \sum_{x \notin Z(G)} |G|/|C(x)|.$$

*Proof* We see immediately that $x$ is a $G$-fixed point if and only if $x \in Z(G)$. Moreover, the stabilizer of any element $x$ is $C(x)$. The formula is now immediate from the Orbit-Stabilizer Formula applied to this specific case.  ∎

**Example 7.2.8** Applying the Orbit-Stabilizer Formula to Example 7.2.3, we see that, on one hand, we have $n^p$ elements in $X$ and, on the other, the set of fixed elements is easily seen to be of size $n$. Every term on the right-hand side sum from Theorem 7.2.5 is $p$ since $\mathbb{Z}_p$ has no non-trivial subgroups, and we obtain Fermat's little theorem again.

## 7.3 Burnside's Lemma

It is possible to derive a formula for the number of equivalence classes under a group action. This is called Burnside's lemma as the English mathematician William Burnside (1852–1927) wrote about it in 1900. The result was known before Burnside, and it appears in the works of the French mathematician Augustin Louis Cauchy (1789–1857) and of the German mathematician Ferdinand Georg Frobenius (1849–1917).

**Theorem 7.3.1** (Burnside's lemma) *If $G$ is a finite group acting on a set $X$, the number of equivalence classes is*

$$\frac{1}{|G|} \sum_{g \in G} |\text{fix}(g)|.$$

*In other words, the number of equivalence classes is the average number of fixed points.*

*Proof* The equivalence class of an element $x$ of $X$ is the orbit of $x$. Thus, if $w(x)$ is $1/|Gx|$, we see that the number of equivalence classes is

$$\sum_{x \in X} w(x).$$

On the other hand, this is

$$\sum_{x \in X} \frac{1}{|G|} |G_x| = \frac{1}{|G|} \sum_{x \in X} \sum_{g \in G_x} 1 = \frac{1}{|G|} \sum_{x \in X} \sum_{g \in G : gx = x} 1.$$

By interchanging the sum, we find this is

$$\frac{1}{|G|} \sum_{g \in G} \sum_{x \in X : gx = x} 1 = \frac{1}{|G|} \sum_{g \in G} |\text{fix}(g)|.$$

This completes the proof. ∎

**Corollary 7.3.2** *The number of conjugacy classes in a group is*

$$\frac{1}{|G|} \sum_{g \in G} |C(g)|.$$

***Proof*** The number of fixed points of $g \in G$ is precisely $|C(g)|$. ∎

**Example 7.3.3** Let us apply this to the problem of counting necklaces. Each necklace of length $n$ formed out of beads of $\lambda$ colours can be viewed as a sequence $(a_1, \ldots, a_n)$ with $a_i \in \{1, 2, \ldots, \lambda\}$. Two necklaces are considered the same if the two sequences representing them are the same after a shift. In other words, $\mathbb{Z}/n\mathbb{Z}$ acts on the sequences and the number of necklaces is precisely the number of equivalence classes under this action. Now, how many fixed points does an element $r$ of $\mathbb{Z}/n\mathbb{Z}$ have A sequence $(a_1, \ldots, a_n)$ is fixed by $r$ if and only if

$$a_{i+tr} = a_i$$

for all $t$ and all $i$. In other words,

$$a_{i+u} = a_i$$

for all $i$ and all $u$ lying in the subgroup generated by $r$ in $\mathbb{Z}/n\mathbb{Z}$. Since $\mathbb{Z}/n\mathbb{Z}$ is cyclic, any subgroup is also cyclic so the number of fixed points of $r$ is $\lambda^{n/o(r)}$ where $o(r)$ is the order of $r$ mod $n$. Recall that in any cyclic group of order $n$, the number of elements of order $d|n$ is precisely $\phi(d)$. Thus, the number of necklaces is

$$\frac{1}{n} \sum_r \lambda^{n/o(r)} = \frac{1}{n} \sum_{d|n} \phi(d) \lambda^{n/d} = \frac{1}{n} \sum_{d|n} \phi(n/d) \lambda^d.$$

## 7.4  Sylow Theorems

It is not our intention to give the student a crash course on group theory. However, several essential ideas will be used repeatedly, and it is helpful if we review them.

First and foremost is the notion of a **quotient group**. If $N$ is a subgroup of a group $G$, we say $N$ is **normal** if $xNx^{-1} = N$ for all $x \in G$. In such a case, the set of cosets of $N$, denoted $G/N$, can be endowed with a group structure in the obvious way: $(aN)(bN) = (ab)N$. It is straightforward to show that this is well defined (and we leave it to the reader as an exercise). The **correspondence theorem** states that there is a one-to-one correspondence between subgroups of $G$ containing $N$ and subgroups of $G/N$. Moreover, normal subgroups correspond to each other. If $G \supseteq H \supseteq N$, the corresponding subgroup of $G/N$ is simply $H/N$. We invite the student to study Chap. 2 of Rotman (1973) for a quick and friendly review of this material.

Let $n$ be a natural number and $p$ a prime dividing $n$. If $G$ is a group of order $n$, consider the set of order $p$-tuples of elements of $G$ whose product is 1:

$$X = \{(x_1, \ldots, x_p) \mid x_1 \cdot \ldots \cdot x_p = 1\}.$$

The size of $X$ is $n^{p-1}$ since we may first choose each of $x_1, \ldots, x_{p-1}$ in $n$ ways, and then $x_p$ is uniquely determined by the equation $x_1 \cdot \ldots \cdot x_p = 1$ as the inverse of $x_1 \cdot \ldots \cdot x_{p-1}$.

Consider the following action of the additive group $(\mathbb{Z}_p, +)$ on $X$ by setting

$$1 \cdot (x_1, \ldots, x_p) = (x_p, x_1, \ldots, x_{p-1}).$$

Note that the set of fixed points consists of elements $(x, x, \ldots, x)$ with $x^p = 1$. Because $p$ is a prime divisor of $n$, the Orbit-Stabilizer Formula gives that the number of fixed points is divisible by $p$. Since $\text{fix}(G) \neq \emptyset$ (why?), it follows that $G$ has an element of order $p$.

**Theorem 7.4.1** (Cauchy, 1845) *If $G$ is a group of order $n$ and $p$ is a prime dividing $n$, then $G$ has an element of order $p$.*

However, much more is true. Cauchy's theorem was generalized by the Norwegian mathematician Peter Ludwig Sylow (1832–1918) in 1872. Almost all work on finite groups use Sylow's theorems. The class equation enables us to deduce the first Sylow theorem, namely:

**Corollary 7.4.2** (Sylow's First Theorem) *If $G$ is a group of order $n$ and $p^k$ is a prime power dividing $n$, then $G$ has a subgroup of order $p^k$.*

*Proof* We proceed by induction on $|G|$. If $|G| = 2$, the theorem is true.

Let $|G| = p^r m$, where $r \geq k$ and $m$ and $p$ are coprime. If $x \in G$ and $p^k$ divides $|C(x)|$, then we are done by induction.

Otherwise, because every term in the sum occurring in the class equation is divisible by $p$, we deduce that $p$ divides the order of the centre $Z(G)$. By Cauchy's theorem, $Z(G)$ has an element $x$ of order $p$. The subgroup generated by $x$ in $G$ is normal since $x \in Z(G)$. The quotient $G/\langle x \rangle$ has order divisible by $p^{k-1}$ and by induction has a subgroup $H/\langle x \rangle$ of order $p^{k-1}$. By the correspondence theorem, $H$ is a subgroup of $G$ of order $p^k$, as desired.                                          ∎

We note that all the Sylow theorems can be derived by considering appropriate group action. Recall the notion of a $p$-Sylow subgroup. If $p^k$ is the largest power of a prime number $p$ dividing the order of $G$, and $P$ is a subgroup of order $p^k$, we call $P$ a $p$-**Sylow subgroup** of $G$. The **normalizer** of a subgroup $H$ of $G$ is $N(H) = \{g : g \in G, gHg^{-1} = H\}$.

**Corollary 7.4.3** (Sylow's Second Theorem) *Let $G$ be a finite group of order $n$ and $P$ a $p$-Sylow subgroup of $G$. Let $X$ be the set of $p$-Sylow subgroups of $G$ and let $P$ act on $X$ via conjugation. Then, $P$ is the only fixed point under this action. Thus, the number of $p$-Sylow subgroups is $\equiv 1(\mod p)$ and all the $p$-Sylow subgroups are conjugates of $P$. Moreover, any $p$-subgroup of $G$ is contained in some conjugate of $P$.*

**Proof** Suppose $Q$ is another $p$-Sylow subgroup fixed by $P$. Then, $gQg^{-1} = Q$ for all $g \in P$. Take $x \in P\backslash Q$. Then, $x$ is in the normalizer $N(Q)$. But $N(Q)$ contains $Q$ and the coset $xQ$ is not $Q$. As the quotient $N(Q)/Q$ has order coprime to $p$, the coset $xQ$ has order $k$ coprime to $p$. Thus, for some $k$, $x^k \in Q$ with $(k, p) = 1$. But $x$ has order equal to some prime power $p^b$ (say). So we can find integers $u, v$ so that $ku + p^b v = 1$. Hence, $x = x^{ku+p^b v} \in Q$, contrary to hypothesis. As the set $X$ is partitioned into orbits under the action of $P$, we deduce immediately that the number of elements of $X$ is $\equiv 1 \pmod{p}$. Now let $Y$ be the set of conjugates of $P$. Let $H$ be a $p$-subgroup of $G$. Then $H$ acts on $Y$. If $H$ fixes an element $Q$ of $Y$, then $H$ is in the normalizer of $Q$. If $H$ is not contained in $Q$, then the argument above gives us a contradiction. Thus, every $p$-subgroup $H$ is contained in some conjugate of $P$. In particular, if $H$ is another $p$-Sylow subgroup, this means that it is conjugate to $P$. This completes the proof.                                          ∎

A $p$-**group** is a group whose order is a power of $p$ where $p$ is a prime number. We observe here that any $p$-group $G$ has subgroups of all orders dividing $|G|$. Indeed, the class equation implies the non-triviality of the centre. By Cauchy's theorem, we may take an element $z$ in the centre of order $p$ and consider the quotient $G/\langle z \rangle$. By induction, this has subgroups of all orders dividing $|G|/p$ which by the correspondence theorem give subgroups of the required order in $G$. For an arbitrary group $G$, and any prime power $p^t$ dividing $|G|$, we deduce that $G$ has subgroups of order $p^t$. Moreover, one can show that the number of these subgroups is $\equiv 1(\mod p)$, but we leave this as an exercise.

Given a finite group $G$ of order $n$, and a subgroup $H$ of $G$, we can partition $G$ into the cosets of $H$ from which we can deduce the famous **Lagrange's theorem**, namely, that the order of the subgroup $H$ divides the order of $G$. The converse is not true, as is seen by considering the alternating group $A_4$ on 4 letters. These are the even permutations of $S_4$ and one can list the elements:

$$(1), (1\,2)(3\,4), (1\,3)(2\,4), (1\,4)(2\,3), (1\,2\,3), (1\,3\,2), (1\,2\,4), (1\,4\,2)$$

$$(2\,3\,4), (2\,4\,3), (3\,4\,1), (3\,1\,4).$$

If $A_4$ had a subgroup $H$ of order 6, then this subgroup is necessarily normal which means that the square of any element of $A_4$ lies in $H$. In particular, the square of any 3-cycle $g$ is in $H$. But $g = (g^2)^2$ lies in $H$ so that all 3-cycles must lie in $H$, a contradiction since there are eight 3-cycles. The virtue of Sylow theory is that it shows that the converse of Lagrange's theorem holds for prime powers dividing the order of the group.

## 7.5  Pólya Theory

George Pólya (1887–1985) was a Hungarian-American mathematician with important contributions to combinatorics, number theory, probability, and mathematical education. His book *How to Solve It* Pólya (1957) conveys to the mathematical novice the nature of mathematical discovery using simple illustrations. We will now describe an important combinatorial theory discovered by Pólya.

The action of a group $G$ on a set $X$ can be viewed as a map

$$G \to \mathrm{Sym}(X),$$

where we send each element $g \in G$ to the permutation $x \mapsto gx$ since $gx = gy$ implies $x = y$ by the axioms of action. In this way, we may view each element of $G$ as a permutation, and so we can consider its cycle decomposition as a product of disjoint cycles. Suppose $g$ has $c_1$ cycles of length 1, $c_2$ cycles of length 2,..., and $c_n$ cycles of length $n$ where $n = |X|$. The **cycle index** of $g$ is defined to be the monomial

$$x_1^{c_1} x_2^{c_2} \cdots x_n^{c_n},$$

which we symbolically denote by $x^g$. The **cycle index polynomial** of $G$ is defined to be the polynomial

$$P_G(x) = \frac{1}{|G|} \sum_{g \in G} x^g.$$

The situation can be looked at in another way. If $G$ acts on $X$ and we have a map $f : X \to Y$, we may view $Y$ as a set of **colours**. Then, the action of $G$ on $X$ induces an action of $G$ on $\mathrm{Map}(X; Y)$, the set of maps from $X$ to $Y$ as follows:

$$(g \cdot f)(x) = f(g^{-1}x).$$

It is important to check that this is indeed an action: we have for $x \in X$,

$$[(gh)f](x) = f((gh)^{-1}x) = f(h^{-1}g^{-1}x).$$

On the other hand,

$$[g(hf)](x) = (hf)(g^{-1}x) = f(h^{-1}g^{-1}x),$$

as desired. Burnside's lemma immediately implies the following.

**Theorem 7.5.1** (Pólya) *Let $X$ and $Y$ be finite sets and $G$ act on $X$. The number of orbits of $G$ on $\text{Map}(X; Y)$ is*

$$\frac{1}{|G|} \sum_{k=1}^{\infty} c_k(G)|Y|^k,$$

*where $c_k(G)$ is the number of elements of $G$ with exactly $k$ disjoint cycles in their cycle decomposition.*

Notice that this number is simply $P_G(|Y|, |Y|, \ldots)$.

**Proof** To apply Burnside's lemma, we must count the number of fixed points of an element $g$ on $\text{Map}(X; Y)$. That is, we must count the number of maps $f : X \to Y$ such that $gf = f$. This means that $f$ is constant on each orbit of $g$. The number of orbits is the number of disjoint cycles in the cycle decomposition of $g$. We may assign values of $f$ arbitrarily on each orbit, so the final count is given as stated in the theorem.                                                                        ∎

If we let $Y$ denote the set of $\lambda$ colours of beads, and $X$ denotes the set $\{1, 2, \ldots, n\}$, then a sequence $(a_1, \ldots, a_n)$ of length $n$ can be viewed as a map $f$ from $X$ to $Y$. As the group $\mathbb{Z}/n\mathbb{Z}$ acts on the coordinates in the obvious way by shifting, this induces an action on $\text{Map}(X; Y)$. We see then that the maps that correspond to distinct necklaces are equivalence classes of maps under this induced action.

We can retrieve our result about the necklace count from the previous section in the following way. First, we must determine the cycle structure of a residue class $r$ viewed as a permutation. Clearly, all orbits have the same length and if $o(r)$ denotes the order of $r$, then each orbit has size $o(r)$ and the number of disjoint cycles is $n/o(r)$. Hence, the number of elements of $\mathbb{Z}/n\mathbb{Z}$ with exactly $k$-cycles is zero unless $k|n$, in which case it is the number of elements of order $n/k$. The number of such elements is $\phi(n/k)$, as we saw before.

Now suppose we have the dihedral group $D_n$ acting on the necklace sequences. Thus, if we present $D_n$ using generators and relations, we can write

$$\langle r, f : r^n = 1, f^2 = 1, frf = r^{-1} \rangle.$$

We could try to count the number of equivalence classes by using Burnside's formula. To use Burnside's formula, we have to count the number of fixed points of each element of $D_n$. It is better to use the cycle index polynomial to determine the number of equivalence classes. We illustrate this as follows.

Firstly, let us have a geometric view of the dihedral group. It is to be viewed as the group of symmetries of a regular $n$-gon. If we fix any vertex, and bisect the interior angle subtended at that vertex, we can view the element $f$ as the flip of the polygon about this axis. We can view the elements $fr^j$ as flips about the axis determined by the other points. If $n$ is odd, each of these elements fixes one vertex and transposes pairs of vertices which are mirror images about that axis. Thus, the cycle structure of $fr^j$ is that it is a product of one-one cycle and $(n-1)/2$ transpositions. Thus, in the case of $n$ odd, the cycle index polynomial is easily seen to be

$$\frac{1}{2n}\left(\sum_{d|n}\phi(d)x_d^{n/d} + nx_1x_2^{(n-1)/2}\right).$$

Now we consider the case $n$ even. As noted above, there are two axes of symmetry. The elements $fr^j$ with $j$ odd correspond to flipping through an axis through a vertex. In this case, it is seen that the opposite vertex is also fixed. In this way, we see the cycle decomposition is a product of $(n-2)/2$ transpositions and two 1-cycles. If $j$ is even, there are no fixed points and the cycle decomposition of $fr^j$ is simply a product of $n/2$ transpositions. In this case, the cycle index polynomial is

$$\frac{1}{2n}\left(\sum_{d|n}\phi(d)x_d^{n/d} + \frac{n}{2}x_1^2x_2^{(n-2)/2} + \frac{n}{2}x_2^{n/2}\right).$$

Pólya's theorem now tells us that the number of equivalence classes of maps is $P_G(\lambda, \lambda, \ldots)$ where $\lambda$ is the number of elements of $Y$.

**Theorem 7.5.2** *Under the action of the dihedral group, the number of distinct necklaces of length $n$ formed using beads of $\lambda$ colours is*

$$\frac{1}{2}\left(\sum_{d|n}\phi(n/d)\lambda^d + \lambda^{(n+1)/2}\right),$$

*if $n$ is odd and*

$$\frac{1}{2}\left(\sum_{d|n}\phi(n/d)\lambda^d + \frac{1}{2}\lambda^{(n+2)/2} + \frac{1}{2}\lambda^{n/2}\right),$$

*if $n$ is even.*

We conclude this section with one application of Pólya theory to chemistry. It seems that the historic origins of the theory are rooted in problems arising in chemistry.

The methane molecule has chemical composition $CH_4$ where $C$ denotes a carbon atom and $H$ is a hydrogen atom. This molecule has tetrahedral shape and the $H_4$

indicates that there are four atoms of hydrogen in the molecule positioned at the vertices of the tetrahedron, with the carbon atom at the centroid. The problem is to determine how many molecules can be formed by replacing the hydrogen atoms with one of bromine, chlorine, or fluorine. This question can be re-interpreted in the context of the colouring problems considered by Pólya theory.

Indeed, the group of symmetries of the regular tetrahedron is $A_4$, the alternating group on four letters. To see this, observe that we can rotate the tetrahedron about the centre of any face and each of these correspond to 3-cycles, one for each face. This gives us a total of eight 3-cycles in the group of symmetries. There is one more symmetry given by a rotation by $180°$ about the axis joining the centre of opposite sides. This is easily seen to be a product of two transpositions and there are three such permutations. Together with the identity, we have the full group of symmetries.

It is now straightforward to write down the cycle index polynomial of the action of $A_4$ on the vertices of the regular tetrahedron. From the discussion above, we have

$$P_{A_4}(x_1, x_2, x_3, x_4) = \frac{1}{12}\left(x_1^4 + 8x_1 x_3 + 3x_2^2\right).$$

The number of different molecules is then seen to be $P_{A_4}(3, 3, 3, 3) = 15$. If the group of symmetries is not taken into account, we have $3^4 = 81$ ways of placing the atoms of bromine, chlorine, or fluorine at the vertices of the tetrahedron. However, many of them give the same molecule.

We make a few additional remarks concerning Pólya's theorem. In the special case that $G = S_n$ acting on the set $\{1, 2, \ldots, n\}$ in the usual way, the cycle index polynomial $P_{S_n}(\lambda, \ldots, \lambda)$ is

$$\frac{1}{n!}\sum_{k=0}^{n} |s(n, k)|\lambda^k,$$

where the $s(n, k)$'s denote the Stirling numbers of the first kind. This represents the number of ways of colouring $n$ indistinguishable objects (or balls) using $\lambda$ colours. This is related to a problem treated earlier by simpler methods. Indeed, this is the same as asking in how many ways we may put $n$ indistinguishable balls into $\lambda$ boxes. This is the same as the number of solutions of

$$x_1 + x_2 + \cdots + x_\lambda = n,$$

with the $x_i$'s non-negative integers. In either interpretation, it is easily seen that the number of ways is $\binom{n+\lambda-1}{\lambda-1}$. Indeed, if we first consider a collection of $n$ distinguishable balls, and we throw into this collection $\lambda - 1$ indistinguishable "sticks", then the number of ways we can arrange these objects is clearly $(n + \lambda - 1)!$. However, $\lambda - 1$ of these objects are identical and can be permuted in $(\lambda - 1)!$ ways, and so we get our result. Now if we say the balls are also indistinguishable, then we can permute these among themselves in $n!$ ways. In this way, we retrieve an earlier formula, namely,

$$(\lambda + n - 1)(\lambda + n - 2) \cdots \lambda = \sum_{k=0}^{n} |s(n,k)| \lambda^k.$$

If we change $\lambda$ to $-\lambda$, we get

$$(\lambda)_n = \sum_{k=0}^{n} s(n,k) \lambda^k.$$

We describe now two further applications of the Pólya theory.

**Example 7.5.3** The game of tic-tac-toe involves a $3 \times 3$ grid in which the players place alternately $\times$ or $\circ$ until a row, column, or diagonal of the same symbols is placed, and the game is over. It is interesting to consider how many possible configurations can be seen at any given moment during a game. Or even, one may ask how many possible outcomes are there. This in its generality is too difficult to answer. We will consider a simpler problem. Namely, in how many ways can we colour a $3 \times 3$ grid using three colours. We can see that the cyclic group of order 4 operates by rotation on such a grid. If we label the grid as

| 1 | 2 | 3 |
|---|---|---|
| 6 | 5 | 4 |
| 7 | 8 | 9 |

then a clockwise rotation $r$ is represented by the permutation

$$(1\,3\,9\,7)(2\,4\,8\,6)(5)$$

whereas $r^2$ is given by

$$(1\,9)(2\,8)(3\,7)(6\,4)(5)$$

$r^3$ has the same cycle structure as $r$ and so we easily see that the cycle index polynomial is

$$P(x_1, \ldots, x_9) = \frac{1}{4} \left( x_1^9 + 2x_1 x_4^2 + x_1 x_2^4 \right).$$

A simple calculation shows that the number of colourings with three colours is 4995. Of course, some of these can never represent the final outcome or the shape of the grid during the game. For such a computation, one needs a finer Pólya theory with weights, which we do not consider here.

**Example 7.5.4** Let us now consider the problem of colouring the faces of the cube using $\lambda$ colours. To do this, we begin by considering the group of symmetries of the cube. These can be classified as follows:

(1) the identity element;
(2) rotation by $90°$ about the axis joining the centre of two opposite faces;

(3)  rotation by 180° about the same axis as in (2);
(4)  rotation by 180° about the axis joining the midpoints of two diagonally opposite edges; and
(5)  rotation by 120° about the axis determined by the diagonal of the cube.

If we think of these symmetries as acting on the faces, and write down the cycle structure, we see that the following:

(1)  1 element of type $1^6$;
(2)  6 elements of type $1^2 4^1$;
(3)  3 elements of type $1^2 2^2$;
(4)  6 elements of type $2^3$; and
(5)  8 elements of type $3^2$.

Thus, we see the group of symmetries has order 24. One can easily see that this group is isomorphic to $S_4$. We can immediately write down the cycle index polynomial for $S_4$ acting on the faces of the cube from the above analysis:

$$P_{S_4}(x_1, \ldots, x_6) = \frac{1}{24}\left(x_1^6 + 6x_1^2 x_4 + 3x_1^2 x_2^2 + 6x_2^3 + 8x_3^2\right).$$

By Pólya's theorem, the number of ways of colouring the faces of the cube using $\lambda$ colours is

$$\frac{1}{24}\left(\lambda^6 + 12\lambda^3 + 3\lambda^4 + 8\lambda^2\right).$$

In particular, there are 10 ways of colouring the faces of the cube using two colours.

## 7.6  Exercises

**Exercise 7.6.1**  Show that the examples from Sect. 7.2 are group actions.

**Exercise 7.6.2**  Let $G$ be a finite group acting on a finite set $X$. For each $g \in G$, define $\sigma_g(x) = g \cdot x$ for each $x \in X$. Show that $\sigma_g$ is a permutation of $X$.

**Exercise 7.6.3**  Let $G$ be a group acting on a set $X$. Show that the map

$$g \mapsto \sigma_g,$$

where $\sigma_g$ is defined as above is a group homomorphism from $G$ into $\mathrm{Sym}(X)$ which is the group of permutations of the set $X$.

**Exercise 7.6.4**  Let $G$ be a group acting on a set $X$ and $H$ a group acting on a set $Y$. Assume that $X$ and $Y$ are disjoint and let $U = X \cup Y$. For $g \in G, h \in H$, define

$$(g, h) \cdot x := g \cdot x \text{ if } x \in X$$

and
$$(g, h) \cdot y := h \cdot y \text{ if } y \in Y.$$

Show that this defines an action of $G \times H$ on $U$.

**Exercise 7.6.5** Determine the number of ways in which four corners of a square can be coloured using two colours. It is permissible to use a single colour on all four corners.

**Exercise 7.6.6** In how many ways can you colour the four corners of a square using three colours?

**Exercise 7.6.7** If $X = \{1, 2, 3\}$, define an action of $S_3$ on $X$ by $\sigma \cdot i = \sigma(i)$ for $i \in X$ and $\sigma \in S_3$. Calculate the cycle index polynomial $P_{S_3}(x_1, x_2, x_3)$.

**Exercise 7.6.8** In how many ways can you colour the vertices of an equilateral triangle with three colours so that at least two colours are used?

**Exercise 7.6.9** What is the number of labelled graphs on four vertices? What is the number of unlabelled graphs on four vertices?

**Exercise 7.6.10** Let $G$ and $H$ be finite groups acting on finite sets $X$ and $Y$. Assume that $X$ and $Y$ are disjoint. By Exercise 7.6.3, we can define an action of $G \times H$ on $X \cup Y$. If $P_G$ and $P_H$ indicate the cycle index polynomials of $G$ acting on $X$ and $H$ acting on $Y$, respectively, show that the cycle index polynomial of $G \times H$ acting on $X \cup Y$ is $P_G P_H$.

**Exercise 7.6.11** How many striped flags are there having six stripes (of equal width) each of which can be coloured red, white, or blue?

**Exercise 7.6.12** What if we change the number of stripes to $n$ and the number of colours to $q$?

**Exercise 7.6.13** Let $n$ be a natural number. Let $S_n$ acting on the set $X = [n]$ in the usual way (such as in Exercise 7.6.7). Let $P_{S_n}$ be the cycle index polynomial. Prove that $P_{S_n}$ is the coefficient of $z^n$ in the power series expansion of

$$\exp(zx_1 + z^2 x_2/2 + z^3 x_3/3 + \dots).$$

**Exercise 7.6.14** For $\sigma \in S_n$, we say $\sigma$ has cycle type $(c_1, c_2, \dots, c_n)$ if $\sigma$ has precisely $c_i$ cycles of length $i$ in its unique decomposition as a product of disjoint cycles. Show that the number of permutations of type $(c_1, c_2, \dots, c_n)$ is

$$\frac{n!}{1^{c_1} c_1! 2^{c_2} c_2! \dots n^{c_n} c_n!}.$$

**Exercise 7.6.15** Let $P_n$ denote the path on $n$ vertices. What is the automorphism group $\text{Aut}(P_n)$ of $P_n$?

**Fig. 7.1** The graph from
Exercise 7.6.18

**Exercise 7.6.16** What is the cycle index polynomial of $\mathrm{Aut}(P_n)$ acting on the vertex set of $P_n$?

**Exercise 7.6.17** In how many ways can we colour the vertices of $P_n$ using $\lambda$ colours, up to the symmetry of $\mathrm{Aut}(P_n)$?

**Exercise 7.6.18** Consider the graph $X$ obtained from the complete graph $K_5$ by deleting two edges incident to the same vertex. What is the automorphism group $\mathrm{Aut}(X)$ of $X$?

**Exercise 7.6.19** Let $X$ be the graph in Fig. 7.1. What is the cycle index polynomial of $\mathrm{Aut}(X)$ acting on the vertex set of $X$?

**Exercise 7.6.20** In how many ways can we colour the vertices of $X$ using $\lambda$ colours, up to the symmetry of $\mathrm{Aut}(X)$?

# References

G. Pólya, *How to Solve It*, 2nd edn. (Doubleday, 1957)
J.J. Rotman, *The Theory of Groups, An Introduction*, 2nd edn. (Allyn and Bacon, Boston, 1973)

# Chapter 8
# Matching Theory

## 8.1 The Marriage Theorem

A **matching** of a graph $X = (V, E)$ is a collection of edges of $X$ which are pairwise disjoint. The vertices incident to the edges of a matching $M$ are **saturated** by $M$. A **perfect matching** is a matching that saturates all the vertices of $X$. Obviously, a necessary condition for that to happen is that $X$ has an even number of vertices, but that is not sufficient in general. Consider the graph $K_{1,3}$, for example (see also Exercise 8.6.8).

Given a bipartite graph $X$ with bipartite sets $A$ and $B$, we would like to know when there is a matching such that each element of $A$ is matched to an element of $B$ uniquely, which is the same as a matching that saturates each vertex of $A$. Thus, a matching is a one-to-one map $f : A \to B$ such that $(a, f(a))$ is an edge of the bipartite graph $X$. This question arises in many *real-life* contexts: $A$ could be a set of jobs a company would like to fill, and $B$ could be a set of candidates applying for the jobs. We would join $a \in A$ to $b \in B$ if $b$ is qualified to do job $a$. The matching question is whether all the jobs can be filled.

This question was formulated in *matrimonial terms* and solved by the English mathematician Philip Hall (1904–1982) in 1935. His theorem goes under the appellation of the *marriage theorem*. Suppose we have a set of $n$ women and $n$ men. We would like to match each woman to a man she likes. Under what conditions can we match all the women? We can encode this information as a bipartite graph $X$, with $A$ being the set of women, $B$ the set of men. We join vertex $a \in A$ to $b \in B$ if $a$ likes $b$. Clearly, for a matching to be possible, each woman must like at least one man. If we have a situation where two women only like one man, then we have a problem and the matching question cannot be solved. More generally, a necessary condition is that for any subset $S$ of $A$, let $N(S)$ be the set of men liked by some women in $S$, then we need $|N(S)| \geq |S|$. Hall's theorem is that this obvious necessary condition is also sufficient. This is one of the simplest, yet powerful, theorems in mathematics with far-reaching applications.

© Hindustan Book Agency 2022
S. M. Cioabă and M. R. Murty, *A First Course in Graph Theory and Combinatorics*,
Texts and Readings in Mathematics 55, https://doi.org/10.1007/978-981-19-0957-3_8

**Theorem 8.1.1** (P. Hall 1933) *Let $X$ be a bipartite graph with partite sets $A$ and $B$. There exists a matching that saturates each vertex of $A$ if and only if for every subset $S$ of $A$,*

$$|N(S)| \geq |S|, \tag{8.1.1}$$

*where $N(S) = \{y \in B : \exists x \in S, x \sim y\}$.*

**Proof** We name (8.1.1) as Hall's condition. The proof is by strong induction on the number of vertices in $A$. The base case $|A| = 1$ is true since, in this case, Hall's condition implies that $X$ has at least one edge incident with the vertex in $A$. Therefore, there is a matching of size one that saturates the vertex of $A$.

For the induction step, let $X$ be a bipartite graph satisfying Hall's condition with $|A| \geq 2$. First, suppose that

$$|N(S)| \geq |S| + 1, \tag{8.1.2}$$

for every proper subset $S$ of $A$. Let $ab$ be an edge in $X$ with $a \in A, b \in B$. Denote by $Y$ the graph obtained by deleting the vertices $a$ and $b$ from $X$. The graph $Y$ is a bipartite graph with partite sets $A' = A \setminus \{a\}$ and $B' = B \setminus \{b\}$. For the graph $Y$, every subset $S$ of $A'$ satisfies Hall's condition $|N(S)| \geq |S|$. Because $|A'| < |A|$, the induction hypothesis implies that $Y$ has a matching that saturates $A'$. Any such matching together with the edge $ab$ is a matching in $X$ that saturates $A$. This finishes the proof of this case.

If condition (8.1.2) is not satisfied for all proper subsets, then for some proper subset $S_0 \subset A$, we must have that

$$|N(S_0)| = |S_0|.$$

Denote by $X_0$ the subgraph of $X$ induced by the partite sets $S_0$ and $N(S_0)$. It is not too difficult to see that $X_0$ satisfies Hall's condition (8.1.1). By the induction hypothesis, the graph $X_0$ contains a matching $M_0$ that saturates $S_0$.

Denote by $X_1$ the subgraph of $X$ induced by the partite sets $A \setminus S_0$ and $B \setminus N(S_0)$. Note that the graph $X_1$ also satisfies Hall's condition. This is because if there were some subset $C \subseteq A \setminus S_0$ such that

$$|N_{X_1}(C)| < |C|,$$

(where the notation $N_{X_1}(C)$ refers to the neighbours of $C$ in $X_1$) then

$$|N_X(S_0 \cup C)| \leq |N_X(S_0)| + |N_{X_1}(C)| < |S_0| + |C| = |S_0 \cup C|,$$

contrary to Hall's condition. Again, by the induction hypothesis, the graph $X_1$ contains a matching $M_1$ that saturates $A \setminus S_0$. The union $M_0 \cup M_1$ is a matching that saturates $A$. This completes the proof. ∎

Let $n$ be a natural number. Suppose that $A_1, \ldots, A_n$ are finite sets. When is it possible to choose $n$ distinct elements $a_1, \ldots, a_n$ with $a_i \in A_i$ for each $1 \leq i \leq n$?

If that is possible, we call $a_1, \ldots, a_n$ **a system of distinct representatives** or an **SDR** of the sets $A_1, \ldots, A_n$. The answer to the above question is actually equivalent to the marriage theorem from the previous section.

**Theorem 8.1.2** *Let $n$ be a natural number and $A_1, \ldots, A_n$ be finite sets. There exists an SDR for $A_1, \ldots, A_n$ if and only if*

$$|\cup_{i \in I} A_i| \geq |I|,$$

*for every subset $I$ of $\{1, \ldots, n\}$.*

**Proof** Consider the bipartite graph $X$ with partite sets $A$ and $B$. The vertices of $A$ correspond to the subsets $A_i$ ($1 \leq i \leq n$) and the vertices of $B$ are the elements of $\cup_{i=1}^{n} A_i$. We join $A_i$ in $A$ to a vertex $a_j \in B$ if and only if $a_j \in A_i$. Choosing a set of distinct representatives is equivalent to finding a matching in $X$ and the condition of the theorem is precisely Hall's condition. Hence, this theorem is proved and is equivalent to Hall's marriage theorem from the previous section. ∎

**Corollary 8.1.3** *Let $k \geq 1$ be a natural number. If $X$ is a bipartite $k$-regular graph with partite sets $A$ and $B$, then $|A| = |B|$ and $X$ contains a perfect matching.*

**Proof** Counting the number of edges of $X$, we get that $k|A| = k|B|$. Because $k \geq 1$, it follows that $|A| = |B|$.

We check that Hall's condition (8.1.1) is satisfied for $X$. Let $S \subseteq A$ and $N(S)$ be its neighbourhood. The number of edges with one endpoint in $S$ and the other in $N(S)$ equals $k|S|$. The number of edges with one endpoint in $N(S)$ and the other in $N(N(S))$ equals $k|N(S)|$. Observe now that $S \subseteq N(N(S))$ and therefore $k|S| \leq k|N(S)|$. Using $k \geq 1$ again, we get that $|S| \leq |N(S)|$. By Hall's marriage theorem, this implies that $X$ has a matching that saturates $A$. Because $|A| = |B|$, this must be a perfect matching. ∎

**Example 8.1.4** At a party, if every man knows $k$ women and every woman knows $k$ men, then it is possible to match every man with a woman he knows.

Consider two collections of finite sets $A_1, \ldots, A_n$ and $B_1, \ldots, B_n$. We consider the problem of when we can find one SDR for both these families. A set of elements $x_1, \ldots, x_n$ is said to be a **system of common representatives** for $A_1, \ldots, A_n$ and $B_1, \ldots, B_n$ if there is a permutation $\sigma \in S_n$ such that $\{x_1, \ldots, x_n\}$ is an SDR for both $A_1, \ldots, A_n$ and $B_{\sigma(1)}, \ldots, B_{\sigma(n)}$.

**Theorem 8.1.5** *Let $n \geq 1$ be a natural number. Assume that $A_1, \ldots, A_n$ and $B_1, \ldots, B_n$ are two collections of finite sets. There exists a system of common representatives for $A_1, \ldots, A_n$ and $B_1, \ldots, B_n$ if and only if for any $k$, the union of any $k$ sets from $A_1, \ldots, A_n$ intersects at least $k$ sets from $B_1, \ldots, B_n$.*

**Proof** We construct a bipartite graph $X$ in which the partite set $A$ consists of the vertices $A_1, \ldots, A_n$ and the set $B$ is formed by the sets $B_1, \ldots, B_n$. We join $A_i$ to

$B_j$ if and only if $A_i \cap B_j \neq \emptyset$. Clearly, the existence of a matching that saturates $A$ is equivalent to the existence of a system of common representatives. The condition of the theorem is precisely Hall's condition. ∎

**Theorem 8.1.6** *Let $G$ be a finite group. If $H$ is a subgroup of $G$ such that $|H| = |G|/r$, then there exist elements $x_1, \ldots, x_r$ in $G$ such that*

$$G = Hx_1 \cup Hx_2 \cup \ldots \cup Hx_r = x_1H \cup x_2H \cup \ldots \cup x_rH.$$

**Proof** We apply Theorem 8.1.5 with the $A_i$'s being the right cosets of $H$ and the $B_j$'s being the left cosets of $H$. Since these cosets are disjoint, the condition of Theorem 8.1.5 is clearly satisfied simply by a cardinality count. The result follows by applying the previous theorem.

## 8.2 Latin Squares

Let $n$ be a natural number. A **Latin square** is a $n \times n$ array with $n$ symbols such that every symbol appears in each row and each column exactly once. The multiplication table for a finite group of order $n$ in an example of a Latin square. However, not all Latin squares are of this form (see the one in Fig. 8.1, for example).

It is not hard to see that for each $n \geq 1$, there exist a Latin square of order $n$. For two natural numbers $r \leq n$, a $r \times n$ **Latin rectangle** is a $r \times n$ matrix with $n$ symbols such that every symbol appears once in each row and at most once in each column. The first $r$ rows of a Latin square form a $r \times n$ Latin rectangle. A natural question is to determine if given a given $r \times n$ Latin rectangle that uses the symbols $\{1, 2, \ldots, n\}$, it is possible to complete it to give a Latin square. The repeated application of the following result shows that the answer is affirmative.

**Proposition 8.2.1** *For any natural numbers $r < n$, any $r \times n$ Latin rectangle can be completed to a $(r + 1) \times n$ Latin rectangle.*

**Proof** Let $R$ be a $r \times n$ Latin rectangle whose entries are from the set $\{1, \ldots, n\}$. We construct a bipartite graph $X$ as follows. For $1 \leq i \leq n$, let $A_i$ be the set of elements of $[n]$ **not** used in the $i$th column. For the Latin rectangle in Fig. 8.2, $A_1 = \{1, 4, 5\}, A_2 = \{1, 3, 4\}$ and the reader can figure out $A_3, A_4, A_5, A_6$. The sets $A_1, \ldots, A_n$ form one partite set of the bipartite graph $X$ while the other partite set consists of the elements $1, \ldots, n$. For $1 \leq i, j \leq n$, we join $A_i$ to $j$ if and only if

**Fig. 8.1** A Latin square of
order 5

| 1 | 2 | 3 | 4 | 5 |
|---|---|---|---|---|
| 2 | 1 | 5 | 3 | 4 |
| 3 | 4 | 2 | 5 | 1 |
| 4 | 5 | 1 | 2 | 3 |
| 5 | 3 | 4 | 1 | 2 |

**Fig. 8.2** A $3 \times 6$ Latin
rectangle

| 3 | 6 | 5 | 2 | 1 | 4 |
|---|---|---|---|---|---|
| 6 | 2 | 4 | 3 | 5 | 1 |
| 2 | 5 | 1 | 6 | 4 | 3 |

$j \in A_i$. For the Latin rectangle in Fig. 8.2, 3 is adjacent to $A_2, A_3, A_4$, for example.
Clearly, the degree of each $A_i$ is $n - r$ since $|A_i| = n - r$. On the other hand, each
element $j$ appears in each row of $R$, and therefore appears in exactly $r$ columns of $R$.
This means that $j$ is not contained, and therefore available for exactly $n - r$ columns
of $R$. Hence, the degree of $j$ in $R$ is $(n - r)$. Thus, the graph $X$ is bipartite regular
with valency $n - r > 0$. By Corollary 8.1.3, it must contain a perfect matching. This
means that there is a way to add a row to $R$ such that the new row is a permutation
of $\{1, \ldots, n\}$ and for each $1 \leq i \leq n$, the entry added in column $i$ is new, meaning
it is not equal to any of the entries of the $i$th column of $R$. This finishes our proof.
∎

Let $n$ be a natural number. A pair of $n \times n$ Latin squares $A = (a_{ij})_{1 \leq i, j \leq n}$ and
$B = (b_{ij})_{1 \leq i, j \leq n}$ are called **orthogonal** if the $n^2$ ordered pairs

$$(a_{ij}, b_{ij}), 1 \leq i, j \leq n$$

are all distinct.

**Example 8.2.2** The following $2 \times 2$ Latin squares

$$\begin{bmatrix} 1 & 2 \\ 2 & 1 \end{bmatrix}, \begin{bmatrix} 2 & 1 \\ 1 & 2 \end{bmatrix}$$

are not orthogonal since the collection of ordered pairs matrix

$$\begin{bmatrix} (1, 2) & (2, 1) \\ (2, 1) & (1, 2) \end{bmatrix}$$

contains repeated pairs.

**Example 8.2.3** The $3 \times 3$ Latin squares below

$$\begin{bmatrix} 1 & 2 & 3 \\ 2 & 3 & 1 \\ 3 & 1 & 2 \end{bmatrix}, \begin{bmatrix} 1 & 2 & 3 \\ 3 & 1 & 2 \\ 2 & 3 & 1 \end{bmatrix}$$

are orthogonal because the nine ordered pairs

$$\begin{bmatrix} (1, 1) & (2, 2) & (3, 3) \\ (2, 3) & (3, 1) & (1, 2) \\ (3, 2) & (1, 3) & (2, 1) \end{bmatrix}$$

are all distinct.

Leonhard Euler (1707–1783) was a Swiss mathematician who made fundamental contributions to many areas of mathematics including graph theory, combinatorics, analysis, and number theory. In 1782, Euler showed how to construct $n \times n$ orthogonal Latin squares when $n$ is odd (see Exercise 8.6.2) or divisible by 4 and conjectured that one cannot construct a pair of orthogonal Latin squares for all $n \equiv 2 \pmod 4$. The case $n = 6$ is known as the **36 officers problem**. It asks if it is possible to arrange 6 regiments of 6 officers each of different ranks in a $6 \times 6$ square so that no rank or regiment will be repeated in a row or column. In 1900, the French mathematician Gaston Tarry (1843–1913) provided a formal proof that there are no orthogonal Latin squares of order 6. In 1960, Raj Chandra Bose (1901–1987), Sharadchandra Shankar Shrikhande (1917–2020), and Ernest Tilden Parker (1926–1991) showed that Euler's conjecture is false for $n > 6$. This means that $n \times n$ orthogonal Latin squares exist for all $n$, except when $n \in \{1, 2, 6\}$.

## 8.3   Doubly Stochastic Matrices

Let $n$ be a natural number. A real $n \times n$ matrix $A = (a_{ij})_{1 \le i, j \le n}$ with non-negative entries is called **doubly stochastic** if the entries in every row sum to 1 and the entries in every column sum to 1. Such matrices arise naturally in probability theory.

**Example 8.3.1** The following is a $4 \times 4$ doubly stochastic matrix:

$$\begin{bmatrix} 0.2 & 0.3 & 0.4 & 0.1 \\ 0.6 & 0.3 & 0 & 0.1 \\ 0 & 0.2 & 0.5 & 0.3 \\ 0.2 & 0.2 & 0.1 & 0.5 \end{bmatrix}.$$

A **permutation matrix** is a doubly stochastic matrix in which every entry is 0 or 1. Every row and every column of a permutation matrix contains a single 1 and the rest of the entries are zero. The set of $n \times n$ permutation matrices forms a group isomorphic to the symmetric group on permutations on $n$ letters. The following result is known as the Birkhoff-von Neumann theorem. It was obtained independently by the American mathematician Garrett Birkhoff (1911–1996) in 1946 (see Birkhoff (1946)), and by the Hungarian-American mathematician John von Neumann in 1953 (see von Neumann (1953)).

**Theorem 8.3.2** (Birkhoff, von Neumann 1946/53) *Every doubly stochastic matrix can be written as a linear combination of permutation matrices.*

***Proof*** We use strong induction on the number of non-zero entries in the matrix. Let $n$ be a natural number. For the base case, the smallest number of non-zero entries of a doubly stochastic $n \times n$ matrix must be $n$ since there is at least one non-zero entry in each row and each column. A doubly stochastic matrix with exactly $n$ non-zero entries must be a permutation matrix, so the base case is true.

For the induction step, let $M = (a_{ij})_{1 \le i,j \le n}$ be a $n \times n$ doubly stochastic matrix with more than $n + 1$ non-zero entries. We define a bipartite graph $X$ with partite sets $A$ and $B$. The vertices of $A$ will be the rows $R_1, \ldots, R_n$ of $A$ and the vertices of $B$ will be the columns $C_1, \ldots, C_n$ of $A$. We join a row $R_i$ to a column $C_j$ if $a_{ij} > 0$. We claim that this bipartite graph satisfies Hall's condition (8.1.1). Indeed, suppose that $|N(S)| < |S|$ for some subset $S$ of $A$. What does this mean? Let $|S| = s$. There are $s$ rows with fewer than $s$ neighbours. If we list our rows horizontally, the neighbours are precisely the columns in which the rows have non-zero entries. Adding up all the entries of each row in the set $S$ gives a total of $s$. Doing the same column-wise gives us at most $|N(S)| < s$, which is a contradiction. Thus, Hall's condition is satisfied and there is a matching that saturates $A$ which is a perfect matching since $|A| = |B| = n$. The existence of a matching means we may select $n$ non-zero entries of $M$ in such a way that each row and each column contains exactly one of them. Of all these non-zero entries, let $c_1$ be one of the least values. Thus, we can write

$$M = c_1 P_1 + R,$$

where $P_1$ is a permutation matrix. Moreover, $(1 - c_1)^{-1} R$ is again a doubly stochastic matrix but with one less non-zero entry. Thus, the proof is completed by inducting on the number of non-zero entries. ∎

For example, for the matrix in Example 8.3.1, in the bipartite graph of rows and columns, $R_2$ is not adjacent to $C_3$ and $R_3$ is not adjacent to $C_1$. In the first step, one can take the perfect matching $R_2 C_1$, $R_1 C_2$, $R_3 C_3$, $R_4 C_4$ for which $c_1 = 0.3$ and

$$P_1 = \begin{bmatrix} 0 & 1 & 0 & 0 \\ 1 & 0 & 0 & 0 \\ 0 & 0 & 1 & 0 \\ 0 & 0 & 0 & 1 \end{bmatrix}, \text{ and } M - c_1 P_1 = \begin{bmatrix} 0.2 & 0 & 0.4 & 0.1 \\ 0.3 & 0.3 & 0 & 0.1 \\ 0 & 0.2 & 0.2 & 0.3 \\ 0.2 & 0.2 & 0.1 & 0.2 \end{bmatrix}.$$

The interested reader may continue the process and write the matrix from Example 8.3.1 as a linear combination of permutation matrices.

## 8.4  Weighted Bipartite Matching

We describe now another application of Hall's marriage theorem to an optimization problem called **the assignment problem**. For some natural number $n$, consider a weighted bipartite graph $K_{n,n}$ with non-negative weights $w_{ij}$ corresponding to the edges $\{i, j\}$ for $1 \le i, j \le n$. The goal is to find a **maximum weight matching**, that is, a perfect matching so that the sum of the weights of the edges in the matching is maximum among all perfect matchings of $K_{n,n}$. For the sake of simplicity, we assume that the weights are non-negative integers, which is usually not a restriction in practice.

The algorithm to find a maximum weight matching that we now describe is called the **Hungarian algorithm**. It was first discovered by the American mathematician Harold Kuhn in 1955 and later revised by another American mathematician James Munkres in 1957. The algorithm is based on the work of the Hungarian mathematicians Dénes König (1884–1944) and Jenö Egerváry (1891–1958) and it was named the Hungarian algorithm by Kuhn in their honour.

Let $W = (w_{ij})_{1 \leq i, j \leq n}$ be the $n \times n$ weight matrix. A perfect matching $M$ in $K_{n,n}$ corresponds to a **transversal** in $W$ which is a set of $n$ entries in $W$ such that no two are in the same row or in the same column. **The weight of the matching/transversal** $M$ is defined as the sum of the weights of its edges/entries. The goal of finding a maximum weight matching is facilitated by supplementary *weights*. We say a collection of numbers $u_1, \ldots, u_n$ and $v_1, \ldots, v_n$ is a **weighted cover** for $W$ if

$$w_{ij} \leq u_i + v_j \quad \forall \ 1 \leq i, j \leq n.$$

The **cost** of a cover $(u, v)$ is defined as

$$c(u, v) := \sum_{i=1}^{n} u_i + \sum_{j=1}^{n} v_j.$$

**Lemma 8.4.1** *For any matching $M$ and any weighted cover $(u, v)$, we have that*

$$w(M) \leq c(u, v).$$

*Moreover, $c(u, v) = w(M)$ if and only if $M$ is a maximum weight matching.*

*Proof* The first part of the lemma is clear simply by summing over all the edges of the matching $M$, the inequality

$$w_{ij} \leq u_i + v_j.$$

Thus, there is no matching with weight greater than $c(u, v)$ for any cover, and the maximal weight is at most the minimal cost of a cover. If $c(u, v) = w(M)$, then we must have the equality

$$w_{ij} = u_i + v_j$$

for all edges $ij$ of the matching $M$ and this must be a matching of maximum weight by what we said above. ∎

This lemma is the basis of the Hungarian algorithm. As mentioned before, we suppose $w_{ij}$ are non-negative integers and this is not any stringent restriction. We begin by choosing an arbitrary cover, which can easily be done simply by choosing $u_i$ to be the largest weight in the $i$th row and $v_i$ to be zero. Clearly,

$$w_{ij} \leq u_i + v_j$$

is satisfied with this choice. Next, we form a bipartite graph $X_{u,v} = (A, B)$ where the vertices of $A$ are the rows of the matrix $W$ and the vertices of $B$ are the columns. We join row $i$ to column $j$ if and only if $w_{ij} = u_i + v_j$. If we have a perfect matching in this graph, we are done by the lemma. Otherwise, Hall's condition is not satisfied and so there is a set of $m$ rows "adjacent" to fewer than $m$ columns. If for each of these rows, we decrease $u_i$ by 1 and increase $v_j$ by 1, and thus get a new sequence $u'_1, ..., u'_n$ and $v'_1, ..., v'_n$, the inequality

$$w_{ij} \leq u'_i + v'_j$$

is satisfied. To see this, note that if $i, j$ are not related, this is clear since we have the strict inequality $w_{ij} < u_i + v_j$. If $i, j$ are related, then the sum $u_i + v_j$ has not changed. We have thus obtained a new cover whose cost is smaller than the earlier one simply because of Hall's condition is violated. The claim is that this converges to the minimal cost and thus the maximal weight transversal. This is clear since we must arrive at a matching for, otherwise we can lower the cost of the cover and this cannot go on endlessly.

To see how to work this algorithm in practice, it is best to use matrices. We illustrate this to determine a maximal transversal in the matrix

$$\begin{bmatrix} 3 & 1 & 4 & 4 & 5 \\ 1 & 4 & 3 & 5 & 4 \\ 7 & 6 & 8 & 7 & 2 \\ 2 & 1 & 3 & 4 & 5 \\ 6 & 3 & 2 & 8 & 7 \end{bmatrix}.$$

We will write the cost covers above the columns and along the rows. The initial cost is obtained by simply taking the largest weight in each row. We write the matrix whose entries are $u_i + v_j - w_{ij}$ alongside:

$$\begin{array}{c} \phantom{5}\ \ 0\ \ 0\ \ 0\ \ 0\ \ 0 \\ \begin{array}{c} 5 \\ 5 \\ 8 \\ 5 \\ 8 \end{array} \begin{bmatrix} 2 & 4 & 1 & 1 & 0 \\ 4 & 1 & 2 & 0 & 1 \\ 1 & 2 & 0 & 1 & 6 \\ 3 & 4 & 2 & 1 & 0 \\ 2 & 5 & 6 & 0 & 1 \end{bmatrix} \end{array}$$

This gives rise to the *equality graph* which is shown in Fig. 8.3.

This bipartite graph does not contain a perfect matching because for $S = \{R_1, R_2, R_3, R_4, R_5\}$ we have that $N(S) = \{C_3, C_4, C_5\}$ and $|N(S)| < |S|$ violating

**Fig. 8.3** An equality graph
in the Hungarian algorithm

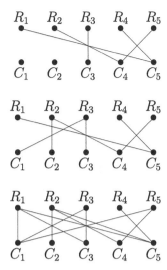

**Fig. 8.4** The second
equality graph in the
Hungarian algorithm

**Fig. 8.5** The third equality
graph in the Hungarian
algorithm

Hall's condition (8.1.1). We decrease each $u_1, u_2, u_3, u_4, u_5$ by 1, and we increase
each $v_3, v_4, v_5$ by 1. This means that the cost of the weight $(u, v)$ goes down by 2.
The updated matrix with entries is $u_i + v_j - w_{ij}$ given by this new cover:

$$
\begin{array}{c}
\phantom{4}\;0\;\;0\;\;1\;\;1\;\;1 \\
\begin{array}{c} 4 \\ 4 \\ 7 \\ 4 \\ 7 \end{array}
\left[\begin{array}{ccccc}
1 & 3 & 1 & 1 & 0 \\
3 & 0 & 2 & 0 & 1 \\
0 & 1 & 0 & 1 & 6 \\
2 & 3 & 2 & 1 & 0 \\
1 & 4 & 6 & 0 & 1
\end{array}\right]
\end{array}
$$

The new equality subgraph is given in Fig. 8.4. This new bipartite graph still does
not have a perfect matching as the set $T = \{R_1, R_2, R_4, R_5\}$ has the property that
$N(T) = \{C_2, C_4, C_5\}$ violating Hall's condition (8.1.1). We can reduce $u_1, u_2, u_4, u_5$
by 1 and increase the $v_2, v_4, v_5$ by 1. The cost of the weight goes down by 1. The
updated matrix is shown below.

$$
\begin{array}{c}
\phantom{3}\;0\;\;1\;\;1\;\;2\;\;2 \\
\begin{array}{c} 3 \\ 3 \\ 7 \\ 3 \\ 6 \end{array}
\left[\begin{array}{ccccc}
0 & 3 & 0 & 1 & 0 \\
2 & 0 & 1 & 0 & 1 \\
0 & 1 & 0 & 1 & 6 \\
1 & 3 & 1 & 1 & 0 \\
0 & 4 & 5 & 0 & 1
\end{array}\right]
\end{array}
$$

The new equality subgraph is given in Fig. 8.5. The equality graph has a perfect
matching which we indicate by $*$ in the matrix below:

$$\begin{bmatrix} 0* & 3 & 0 & 1 & 0 \\ 2 & 0* & 1 & 0 & 1 \\ 0 & 1 & 0* & 1 & 6 \\ 1 & 3 & 1 & 1 & 0* \\ 0 & 4 & 5 & 0* & 1 \end{bmatrix}$$

We have found a matching/transversal $M$ and a cover $(u, v)$ such that $w(M) = c(u, v)$. Hence, the matching has maximum weight which equals $3 + 4 + 8 + 5 + 8 = 28$.

If we were interested in a minimal transversal, all we need to do is to take the maximum $M$ of all the entries and replace our weights $w_{ij}$ by $M - w_{ij}$ and repeat the above algorithm.

## 8.5  Matchings and Connectivity

In a bipartite graph $X$ with bipartite sets $A$ and $B$, Hall's marriage theorem gives a necessary and sufficient condition for the existence of a matching that saturates $A$. For general graphs, the following theorem gives a necessary and sufficient condition for the existence of a perfect matching. It was proved by William Tutte (1917–2002) in 1947. Tutte was one of the leading mathematicians in graph theory and combinatorics. In 1935, he began his studies at Cambridge in chemistry, but soon after he became interested in mathematics. During World War II, he worked at Bletchley Park as a code breaker, and he was able to deduce the structure of a German encryption machine using only a number of intercepted encrypted messages.

An **odd component** of a graph $Y$ is a component of $Y$ that has an odd number of vertices. Let odd$(Y)$ denote the number of odd components of $Y$.

**Theorem 8.5.1**  (Tutte 1947) *A graph $X$ contains a perfect matching if and only if*

$$\text{odd}(X \setminus S) \leq |S| \tag{8.5.1}$$

*for each $S \subset V$.*

**Proof** If $X$ has a perfect matching and $S$ is a subset of vertices of $X$, then each odd component of $X \setminus S$ has a vertex adjacent to a vertex in $S$. This means odd$(X \setminus S) \geq |S|$.

The proof of sufficiency is more complicated. We start it here and invite the reader to complete it.

Assume that condition (8.5.1) is satisfied for all $S \subset V(X)$. Note that by adding edges to $X$, condition (8.5.1) is preserved (prove this). The theorem is true unless there exists a graph $X$ satisfying Tutte's condition that has no perfect matchings and adding any missing edges would create a graph with a perfect matching.

Let $X$ be such a graph. We will obtain a contradiction by showing that $X$ actually contains a perfect matching. Let $C$ denote the set of vertices whose degree is $|V(X)| - $

1. If $X \setminus C$ is formed by disjoint complete graphs, then one can find a perfect matching easily. The case when $X \setminus C$ is not a union of disjoint cliques is left as an exercise.
∎

Tutte's theorem was extended by the French mathematician Claude Berge (1926–2002) in 1958. Berge was one of the leading mathematicians in graph theory and combinatorics in the last century. His result gives a formula for **the matching number** $v(X)$ of a graph $X$ which is defined as the maximum number of edges of a matching in $X$.

**Theorem 8.5.2** (Berge 1958) *For a graph $X = (V, E)$ with $n$ vertices,*

$$v(X) = \frac{1}{2} \left( n - \max_{S \subset V} (\text{odd}(X \setminus S) - |S|) \right).$$

**The vertex-connectivity** $\kappa(X)$ of a connected non-complete graph $X$ is the minimum number of vertices whose removal disconnects $X$. Because $X$ is connected, $\kappa(X) \geq 1$. If $\kappa(X) = 1$, then any vertex whose removal disconnects $X$ is called a **cut-vertex**.

**The edge-connectivity** $\kappa'(X)$ of a connected non-complete graph $X$ is the minimum number of edges whose removal disconnects $X$. If $\kappa'(X) = 1$, then any edge whose deletion disconnects $X$ is called a **bridge**. For the complete graph $K_n$ with $n \geq 2$, we define $\kappa(K_n) = \kappa'(K_n) = n - 1$. The following inequalities hold in any connected graph.

**Proposition 8.5.3** *If $X$ is a connected graph, then*

$$1 \leq \kappa(X) \leq \kappa'(X) \leq \delta(X),$$

*where $\delta(X)$ denotes the minimum degree of $X$.*

**Proof** We may assume that $X$ is not a complete graph. The inequality $1 \leq \kappa(X)$ was explained earlier. If $x$ is a vertex of degree $\delta(X)$, then deleting all the edges incident with $X$ disconnects $x$ from the rest of the graph. Hence, $\kappa'(X) \leq \delta(X)$.

If $\kappa'(X) = 1$, then the inequality $\kappa(X) \leq \kappa'(X)$ holds. This is because if $e$ is an edge whose deletion disconnects $X$, then there is a way to choose one of the endpoints of $e$ such that the removal of that vertex disconnects the graph (Fig. 8.6).

Assume that $\kappa'(X) \geq 2$. Let $x_1 y_1, \ldots, x_k y_k$ be a set of $k = \kappa'(X)$ edges whose removal disconnects $X$. If removing $\{x_1, \ldots, x_k\}$ also disconnects $X$, then $\kappa(X) \leq k$ and we are done. Otherwise, it means that the degree of each $x_i$ is at most $k$, which implies that $\kappa(X) \leq k$.
∎

A graph $X$ is called $k$-**connected** if $\kappa(X) \geq k$. This means that the deletion of any $k - 1$ vertices of $X$ will not disconnect $X$. Similarly, $X$ is called $k$-**edge-connected** if $\kappa'(X) \geq k$. Two $(x, y)$-paths are called **independent** if they only have $x$ and $y$ in common. The following result gives a characterization for a graph being 2-connected.

**Fig. 8.6** A 4-regular graph
with $\kappa = 1$ and $\kappa' = 2$

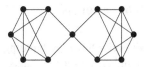

**Theorem 8.5.4** *A graph X is 2-connected if and only if, for any two distinct vertices x and y of X, there is a cycle that contains them or, equivalently, there are at least two independent (x, y)-paths.*

**Proof** Assume that for any two vertices $x$ and $y$, there is a cycle that contains them. Consider a vertex $z$ of $X$. If the removal of $z$ disconnects the graph, let $x$ and $y$ be two vertices in different components of $X - z$. Any $(x, y)$-path must go through $z$ and therefore, there cannot be a cycle containing $x$ and $y$.

Assume that $X$ is 2-connected. Let $x \neq y$ be two vertices of $X$. We use induction on $d(x, y)$ to prove that $x$ and $y$ are on a cycle. If $x$ and $y$ are adjacent, then there must be another $(x, y)$-path of length two or more since, otherwise, $X - xy$ would be disconnected, in contradiction with $X$ being 2-connected. If $d(x, y) \geq 2$, consider a $(x, y)$-path $P$ and the vertex $z$ on it that is adjacent to $y$. By the induction hypothesis, $x$ and $z$ are on a cycle $C$. If that cycle contains $y$, then we are done. Otherwise, because $X$ is 2-connected, $X - z$ must be connected, and therefore it must contain a path $P$ from $x$ to $y$. If the path $P$ intersects $C$ only at $x$, then a $(x, z)$-path from $C$ extended with the edge $zy$ is a $(x, y)$-path that is independent of $P$. Otherwise, if $P$ intersects $C$ in $x$ and other vertices, then we leave it as an exercise for the reader to find two independent $(x, y)$-paths and finish the proof. ∎

A far-reaching generalization of this result, which is the fundamental result involving graph connectivity, was proved in 1927 by the German mathematician Karl Menger (1902–1985). Menger's theorem is an example of a min-max theorem. Given a graph $X$ and two distinct vertices $x$ and $y$ of $X$, let $\kappa(x, y)$ denote the minimum number of vertices whose removal separates $x$ from $y$ (meaning that $x$ and $y$ are in different components of the resulting graph).

**Theorem 8.5.5** *Let $X = (V, E)$ be a connected graph.*

*(1) For any distinct and non-adjacent vertices x and y, $\kappa(x, y)$ equals the maximum number of independent (x, y)-paths.*
*(2) For any $x \neq y \in V$, the minimum number of edges whose removal separates x from y equals the maximum number of edge-disjoint (x, y)-paths.*

**Proof** One inequality is obvious. If there are $r$ independent paths from $x$ to $y$, then deleting one internal vertex from each path will separate $x$ from $y$. The other inequality is more involved and will be omitted here. ∎

## 8.6   Exercises

**Exercise 8.6.1** A building contractor advertises for a bricklayer, a carpenter, a plumber, and a toolmaker; he has five applicants—one for the job of bricklayer, one for the job of carpenter, one for the jobs of bricklayer and plumber, and two for the jobs of plumber and toolmaker. Can the jobs be filled? In how many ways?

**Exercise 8.6.2** Let $n$ be an odd natural number. Consider the following $n \times n$ matrices whose rows, columns, and entries are in the set $\mathbb{Z}_n$ of integers modulo $n$:

$$L_1 : \mathbb{Z}_n \times \mathbb{Z}_n \to \mathbb{Z}_n, L_1(x, y) \equiv x + y \pmod{n}$$
$$L_2 : \mathbb{Z}_n \times \mathbb{Z}_n \to \mathbb{Z}_n, L_2(x, y) \equiv x - y \pmod{n}.$$

Prove that $L_1$ and $L_2$ are orthogonal Latin squares.

**Exercise 8.6.3** A permutation matrix is a 0, 1 matrix having exactly one 1 in each row and column. Prove that a square matrix of non-negative integers can be expressed as a sum of $k$ permutation matrices if and only if all row sums and column sums are equal to $k$.

**Exercise 8.6.4** Let $X = (A, B)$ be a bipartite graph and suppose that $A$ satisfies Hall's condition. Suppose further that each vertex of $A$ is joined to at least $t$ elements of $B$. If $|A| \geq t$, show that the number of matchings that saturate $A$ is at least $t!$.

**Exercise 8.6.5** Show that there are at least $n!(n-1)! \cdots 2!1!$ Latin squares of order $n$. Show that this quantity is greater than $2^{(n-1)^2}$ for $n \geq 5$.

**Exercise 8.6.6** Let $r$ and $s$ be two natural numbers. Consider two partitions of the set $[rs] := \{1, \ldots, rs\}$ into $r$ sets of size $s$:

$$[rs] = A_1 \cup \ldots \cup A_r = B_1 \cup \ldots \cup B_r.$$

Show that there exists a permutation $\sigma \in S_r$ such that

$$A_j \cap B_{\sigma(j)} \neq \emptyset, \forall 1 \leq j \leq r.$$

**Exercise 8.6.7** Find a minimum weight transversal in the matrix below:

$$\begin{bmatrix} 4 & 5 & 8 & 10 & 11 \\ 7 & 6 & 5 & 7 & 4 \\ 8 & 5 & 12 & 9 & 6 \\ 6 & 6 & 13 & 10 & 7 \\ 4 & 5 & 7 & 9 & 8 \end{bmatrix}.$$

**Exercise 8.6.8** Determine whether the graph below has a perfect matching. What is the size of its largest matching?

**Fig. 8.7** The 3-regular
graph for Exercise 8.6.8

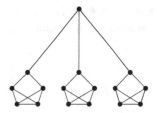

**Exercise 8.6.9** For each $k \geq 2$, construct a $k$-regular graph on an even number of vertices containing no perfect matchings (Fig. 8.7).

**Exercise 8.6.10** Show that in the complete graph $K_{2n}$ the number of perfect matchings is $(2n)!/2^n n!$.

**Exercise 8.6.11** Let $W = (w_{ij})_{1 \leq i,j \leq n}$ a $n \times n$ matrix of non-negative weights. Define a function $f$ on the set of $n \times n$ doubly stochastic matrices by setting for $A = (a_{ij})$,

$$f(A) = \sum_{i,j} a_{ij} w_{ij},$$

where the summation is over all indices $i, j$. Show that $f$ attains its maximum value at a permutation matrix.

**Exercise 8.6.12** Let $t \geq 0$ be an integer. If $X$ is a bipartite graph with partite sets $A$ and $B$ such that $|N(S)| \geq |S| - t$ for each $S \subset A$, then show that $X$ contains a matching that saturates $|A| - t$ vertices of $A$.

**Exercise 8.6.13** Let $t \geq 1$ be an integer. If $X$ is a bipartite graph with bipartite sets $A$ and $B$ such that $|N(S)| \geq t \cdot |S|$ for each $S \subset A$, then each $a \in A$ has a set $S_a$ of $t$ neighbours in $B$ with $S_a \cap S_{a'} = \emptyset$ for each $a \neq a' \in A$.

**Exercise 8.6.14** Let $A$ be a matrix with entries 0 or 1. Show that the minimum number of rows and columns that contain all the 1's of $A$ equals the maximum number of 1's in $A$, no two on the same row or column.

**Exercise 8.6.15** Finish the proof of Theorem 8.5.1.

**Exercise 8.6.16** Show that any connected 3-regular graph with no bridges contains a perfect matching.

**Exercise 8.6.17** Prove that every tree has at most one perfect matching.

**Exercise 8.6.18** Show that a tree $T$ has a perfect matching if and only if $\text{odd}(T \setminus x) = 1$ for any vertex $x$ of $T$.

**Exercise 8.6.19** Let $X$ be a bipartite graph with bipartite sets $A$ and $B$ such that $|N(S)| > |S|$ for each $S \subset A$. Show that for any edge $e$ of $X$, there exists a matching that contains $e$ and saturates $A$.

**Exercise 8.6.20** Let $V_1, \ldots, V_n$ be subspaces of a vector space $V$. Then $V_1, \ldots, V_n$ has a linearly independent system of distinct representatives if and only if

$$\dim(\mathrm{span}(\cup_{i \in I} V_i)) \geq |I|,$$

for each $I \subset [n]$.

# References

G. Birkhoff, Three observations on linear algebra. Univ. Nac. Tucumań. Revista A. **5**, 147–151 (1946)

J. von Neumann, A certain zero-sum two-person game equivalent to the optimal assignment problem, in *Contributions to the Theory of Games*, vol. 2, pp. 5–12, Annals of Mathematics Studies, no. 28 (Princeton University Press, Princeton, N.J., 1953)

# Chapter 9
# Block Designs

## 9.1 Gaussian Binomial Coefficients

CioabăLet $V$ be a $n$-dimensional vector space over the finite field $\mathbb{F}_q$ of $q$ elements. Our first result determines the number $\begin{bmatrix} n \\ k \end{bmatrix}_q$ of $k$-dimensional vector subspaces of $V$.

**Theorem 9.1.1** *Let $V$ be a vector space of dimension $n$ over $\mathbb{F}_q$. The number of $k$-dimensional subspaces in $V$ is*

$$\begin{bmatrix} n \\ k \end{bmatrix}_q = \frac{(q^n - 1)(q^n - q) \ldots (q^n - q^{k-1})}{(q^k - 1)(q^k - q) \ldots (q^k - q^{k-1})}.$$

*Proof* The number of one-dimensional subspaces is easily found, as these are subspaces spanned by one non-zero vector and there are $q^n - 1$ such vectors. But for each choice of a non-zero vector, any non-zero scalar multiple of it will generate the same subspace as there are $q - 1$ such multiples for any fixed vector, we get a final tally of $\frac{q^n-1}{q-1}$. Hence,

$$\begin{bmatrix} n \\ 1 \end{bmatrix}_q = \frac{q^n - 1}{q - 1}.$$

This gives us a clue of how to determine the general formula. Each subspace of dimension $k$ has a basis of $k$ elements. Let us first count in how many ways we can write down a basis for a $k$-dimensional subspace of $V$. For the first vector, we have $q^n - 1$ choices. For the second, we have $q^n - q$ choices, since we must not pick any scalar multiple of the first vector chosen. For the third vector, we have $q^n - q^2$ such vectors, since we should not pick any linear combination of the first two chosen. In this way, we see that the number of ways of writing down a basis for a $k$-dimensional subspace is

$$(q^n - 1)(q^n - q) \ldots (q^n - q^{k-1}).$$

© Hindustan Book Agency 2022
S. M. Cioabă and M. R. Murty, *A First Course in Graph Theory and Combinatorics*,
Texts and Readings in Mathematics 55, https://doi.org/10.1007/978-981-19-0957-3_9

On the other hand, any $k$-dimensional subspace is isomorphic to $\mathbb{F}^k$ and the number of its ordered bases is the number of $k \times k$ non-singular matrices over $\mathbb{F}_q$ which equals

$$(q^k - 1)(q^k - q) \ldots (q^k - q^{k-1}).$$

Combining the previous arguments, we get the desired result.                        ∎

The numbers $\begin{bmatrix} n \\ k \end{bmatrix}_q$ are called $q$-**binomial coefficients** or **Gaussian binomial coefficients**. Observe that if we think of $q$ as a real number and take limits as $q \to 1^+$, we obtain by l'Hôspital's rule that

$$\lim_{q \to 1^+} \begin{bmatrix} n \\ k \end{bmatrix}_q = \binom{n}{k}.$$

The $q$-binomial coefficient numbers have similar properties to the binomial coefficients. This perspective has proved useful in trying to obtain $q$-analogs of classical binomial identities and to understand their meaning from the standpoint of these subspaces.

**Theorem 9.1.2** *If $n$ is a natural number, then*

$$\begin{bmatrix} n \\ k \end{bmatrix}_q = \begin{bmatrix} n \\ n - k \end{bmatrix}_q,$$

*for any integer $k$ with $0 \leq k \leq n$.*

***Proof*** Let $V$ be a vector space of dimension $n$ over $\mathbb{F}_q$. There is a bijection between the collection of $k$-dimensional subspaces of $V$ and the family of $(n - k)$-dimensional subspaces of $V$ given by

$$W \mapsto W^\perp = \{u \in V : u \perp w, \forall w \in W\}.$$

The result above can also be verified directly using Theorem 9.1.1.                  ∎

To any ordered basis, we can associate a $k \times n$ matrix with the basis vectors being the rows. We can view our subspace of dimension $k$ as the row span of this matrix. The row span is unchanged if we perform *row operations* on it as follows. We can multiply any row by a non-zero scalar. We can add one row to another. We can interchange rows. This allows us to speak about **the reduced row echelon form** of a matrix. This form is characterized by the fact that the first non-zero entry of each row is a 1. For any row, all the entries preceding the leading 1 are zero. If a column contains a leading 1, then all its other entries are zero.

**Example 9.1.3** If $n = 4$ and $k = 2$, the possible echelon forms are given by

$$\begin{bmatrix} 1\ 0\ *\ * \\ 0\ 1\ *\ * \end{bmatrix}, \quad \begin{bmatrix} 1\ *\ 0\ * \\ 0\ 0\ 1\ * \end{bmatrix}, \quad \begin{bmatrix} 1\ *\ *\ 0 \\ 0\ 0\ 0\ 1 \end{bmatrix},$$

$$\begin{bmatrix} 0\ 1\ 0\ * \\ 0\ 0\ 1\ * \end{bmatrix}, \quad \begin{bmatrix} 0\ 1\ *\ 0 \\ 0\ 0\ 0\ 1 \end{bmatrix}, \quad \begin{bmatrix} 0\ 0\ 1\ 0 \\ 0\ 0\ 0\ 1 \end{bmatrix},$$

where $*$ denotes any element of $\mathbb{F}_q$. It is clear that every subspace of dimension $k$ has a unique echelon form. Thus, the number of subspaces of dimension $k$ is equal to the number of echelon forms for a $k \times n$ matrix over $\mathbb{F}_q$. In the above example, this number is easily seen to be

$$q^4 + q^3 + 2q^2 + q + 1 = \frac{(q^4 - 1)(q^4 - q)}{(q^2 - 1)(q^2 - q)}.$$

The argument in the previous example can be generalized.

**Theorem 9.1.4** *If $n \geq k$ are two natural numbers, then*

$$\begin{bmatrix} n \\ k \end{bmatrix}_q = \sum_{\ell=0}^{k(n-k)} a_\ell q^\ell,$$

*where $a_\ell$ equals the number of partitions of $\ell$ into $k$ parts of size at most $n - k$.*

We now establish a $q$-analog of Pascal's triangle recurrence relation for the binomial coefficients (see Exercise 2.6.5).

**Theorem 9.1.5** *If $n \geq k$ are two natural numbers, then*

*(1)*

$$\begin{bmatrix} n+1 \\ k \end{bmatrix}_q = \begin{bmatrix} n \\ k-1 \end{bmatrix}_q + q^k \begin{bmatrix} n \\ k \end{bmatrix}_q.$$

*(2)*

$$\begin{bmatrix} n+1 \\ k \end{bmatrix}_q = \begin{bmatrix} n \\ k \end{bmatrix}_q + q^{n+1-k} \begin{bmatrix} n \\ k-1 \end{bmatrix}_q.$$

***Proof*** We prove the first identity by counting the number of reduced row echelon forms. The left-hand side is the number of reduced row echelon forms of a $k \times (n + 1)$ matrix over $\mathbb{F}_q$. Such an echelon form either has a leading 1 in the $(k, n + 1)$-entry or it does not. For those that do, we see that the $(k - 1) \times n$ matrix formed by the first $k - 1$ rows and first $n$ columns is in echelon form and their number is

$$\begin{bmatrix} n \\ k-1 \end{bmatrix}_q.$$

If the $(k, n + 1)$ entry is not a leading 1, then the last column of such a reduced row echelon form has arbitrary entries. The $k \times n$ submatrix obtained by taking the first

$n$ columns is in reduced row echelon form and thus counts the number of subspaces of dimension $k$ in a $n$-dimensional vector space. This number is

$$\begin{bmatrix} n \\ k \end{bmatrix}_q.$$

As we have $q^k$ choices for the last column, we obtain the promised identity. Another proof (less illuminating perhaps) can be obtained by using the formula from Theorem 9.1.5.

Combining Theorem 9.1.5 with Theorem 9.1.2, we deduce the second identity. We leave the details to the reader.                                                   ∎

Note that each of the previous results reduces to the usual recurrence relation for binomial coefficients when $q = 1$. We will now prove the $q$-binomial theorem.

**Theorem 9.1.6** (The $q$-binomial theorem) *If $n$ is a natural number and $t$ a real number, then*

$$\prod_{i=0}^{n-1}(1 + q^i t) = \sum_{k=0}^{n} \begin{bmatrix} n \\ k \end{bmatrix}_q q^{\binom{k}{2}} t^k.$$

***Proof*** We use induction on $n$. For $n = 1$, both sides of the equation are $1 + t$. Let $n$ be a natural number and suppose that the result is true for $n$:

$$\prod_{i=0}^{n-1}(1 + q^i t) = \sum_{k=0}^{n} \begin{bmatrix} n \\ k \end{bmatrix}_q q^{\binom{k}{2}} t^k.$$

Multiplying both sides by $1 + q^n t$, we get that

$$\prod_{i=0}^{n}(1 + q^i t) = (1 + q^n t)\left(\sum_{k=0}^{n} \begin{bmatrix} n \\ k \end{bmatrix}_q q^{\binom{k}{2}} t^k\right).$$

The coefficient of $t^k$ on the right-hand side is

$$q^{\binom{k}{2}} \begin{bmatrix} n \\ k \end{bmatrix}_q + q^{\binom{k-1}{2}} \begin{bmatrix} n \\ k-1 \end{bmatrix}_q q^n$$

which is equal to

$$q^{\binom{k}{2}}\left(\begin{bmatrix} n \\ k \end{bmatrix}_q + q^{n-k+1} \begin{bmatrix} n \\ k-1 \end{bmatrix}_q\right) = q^{\binom{k}{2}} \begin{bmatrix} n+1 \\ k \end{bmatrix}_q,$$

as desired.                                                                      ∎

## 9.2 Design Theory

Design theory has its practical origins in statistics, where one must set up *experiments* or *clinical trials* to test the reliability of a product. Consider the following problem. Suppose that we have 7 volunteers to test 7 products. Each person is willing to test 3 products, and each product should be tested by 3 people to ensure objectivity. Can we arrange the experiment so that any two people would have tested precisely one product in common?

Surprisingly, a solution is provided to this problem by **the Fano plane** (see Fig. 9.1). This name honours the Italian mathematician Gino Fano (1871–1952) who was one of the pioneers of finite geometry.

Consider the triangle of three points; we join each vertex to the midpoint of the opposite side. The three midpoints are then joined by a circle. In this way, we have 7 points and 7 *lines*. Each line would represent a product, and the three vertices on a line would mark out three volunteers to test that particular product. Since any two points determine a unique line, we deduce that any two people test precisely one product in common. Observe that in this situation, any two products are simultaneously tested by precisely one person.

The actual origin of design theory is another famous problem, called **Kirkman's schoolgirls problem**. Thomas Kirkman (1806–1895) was a English mathematician and minister who published this problem in *Lady's and Gentleman's Diary* in 1850.

*Fifteen schoolgirls walk home each day in five groups of three. Is it possible to arrange the walks over a 1-week period so that any two girls walk precisely once together in a group?*

The presentation we give here is adapted from the interesting article Brown and Mellinger (2009). Consider the vector space $\mathbb{F}_2^4$ and remove the zero vector. We then have 15 vectors, and we will use the correspondence between the set $\{1, \ldots, 15\}$ and their binary representations:

**Fig. 9.1** The Fano plane

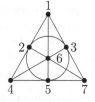

$$1 = 0001$$
$$2 = 0010$$
$$3 = 0011$$
$$\vdots = \vdots$$
$$15 = 1111.$$

Consider triples of vectors $\{x, y, z\}$ such that $x + y + z = 0$. The number of such vectors is 35 since we have 15 choices for $x$, 14 choices for $y$, and then $z$ is uniquely determined. Note that necessarily, the elements of each triple are distinct since if two of them were equal, we get the remaining vector would be zero, which is impossible. The number of ordered triples is $15 \times 14$ and we must divide this number by $3! = 6$ to get 35 unordered triples. Each triple corresponds to a two-dimensional vector space of $\mathbb{F}_2^4$.

It is possible to arrange the solution vectors in 7 groups so that in each group, we have 5 triples and the union of the triples is the set of 15 vectors.

| Mon | Tue | Wed | Thu | Fri | Sat | Sun |
|------|-------|--------|--------|--------|--------|---------|
| 1,2,3 | 1,4,5 | 2,4,6 | 1,6,7 | 3,4,7 | 3,5,6 | 2,5,7 |
| 4,10,4 | 2,13,15 | 1,8,9 | 2,9,11 | 2,12,14 | 2,8,10 | 1,14,15 |
| 7,8,15 | 3,9,10 | 3,12,15 | 4,8,12 | 1,10,11 | 4,11,15 | 4,9,13 |
| 5,9,12 | 6,8,14 | 5,11,14 | 3,13,14 | 5,8,13 | 1,12,13 | 3,8,11 |
| 6,11,13 | 7,11,12 | 7,10,13 | 5,10,15 | 6,9,15 | 7,9,14 | 6,10,12 |

Thus, if we think each schoolgirl corresponding to a vector, this configuration gives us the solution.

To understand precisely what is behind this solution, we must understand the theory of designs. It might be more illuminating to consider the following setup. Let $\mathcal{P}$ be a set of $v$ volunteers or **points**, $\mathcal{B}$ a set of $b$ products or **blocks** as they are called in design theory. We require that each volunteer test $r$ products, and each product should be tested by $k$ people. We can identify a block/experiment with the $k$-subset of points/volunteers corresponding to it. In addition, we require that any pair of people together test precisely $\lambda$ products. Can such an experiment be arranged? The incidence structure of points and blocks with the properties above is called a $2 - (v, k, \lambda)$ design. Sometimes, the more cumbersome notation of a $(b, v, r, k, \lambda)$-design is used. But since $v$, $\lambda$, and $k$ give us $r$ and then $b$ by the theorem below, it is prudent to drop the extra parameters.

**Theorem 9.2.1** *In any* $2 - (v, k, \lambda)$ *design, with $b$ blocks and where each point appears in $r$ blocks, we must have*

$$vr = bk, \quad \text{and} \quad (v - 1)\lambda = (k - 1)r.$$

***Proof*** We can represent this situation by a bipartite graph $X$ with partite sets $\mathcal{P}$ and $\mathcal{B}$, where $\mathcal{P}$ consists of the set of $v$ points and $\mathcal{B}$ is the set of $b$ blocks. We join a point of $\mathcal{P}$ to a block of $\mathcal{B}$ if the point is contained in the block. The conditions tell us that the degree of every vertex in $\mathcal{P}$ is $r$ and the degree of every vertex in $\mathcal{B}$ is $k$. The final condition tells us that any pair of vertices of $\mathcal{P}$ have precisely $\lambda$ common neighbours. We count the number of edges, and we deduce that $vr = bk$.

Now let us construct another bipartite graph in which the partite set on the left consists of all the pairs of points and the partite set on the right is the set of blocks. We join a pair to a block if that pair is contained in that block. This gives $v(v-1)\lambda/2$ edges. Since each block has $k$ elements in it, there are $k(k-1)/2$ pairs that each block will be adjacent to, and so we get $v(v-1)\lambda = k(k-1)b$. Since $vr = bk$, we obtain $(v-1)\lambda = (k-1)r$, which finishes the proof. ∎

These conditions are necessary, but as we shall see below, they are not sufficient. For instance, it will be seen that there is no way to arrange 22 objects into 22 blocks, with each object occurring in precisely 7 blocks and each block containing 7 objects so that any two distinct objects occur in precisely 2 blocks. This corresponds to a $2 - (22, 7, 2)$ design.

A $2 - (v, 3, 1)$ design is called a **Steiner triple system** and denoted by $STS(v)$.

**Example 9.2.2** The Fano plane in Fig. 9.1 is a $2 - (7, 3, 1)$ or $STS(7)$.

**Example 9.2.3** Consider the incidence structure where the points are the 10 edges of the complete graph $K_5$ and the blocks are of three types:

- five stars $K_{1,4}$,
- ten disjoint unions of $K_2$ and $K_3$, and
- five cycles $C_4$.

The motivated reader should prove that this is a $3 - (10, 4, 3)$ design.

A projective plane of order $n$ is a collection $X$ of $n^2 + n + 1$ elements called *points* and a collection $B$ of $n^2 + n + 1$ blocks called *lines*. We require that each point is on precisely $n + 1$ lines and each line has precisely $n + 1$ points, and any two distinct points determine a unique line. Thus, a projective plane of order $n$ is a $2 - (n^2 + n + 1, n + 1, 1)$ design.

**Example 9.2.4** Let $\mathcal{P}$ be the collection of one-dimensional subspaces of $\mathbb{F}_q^3$. The number of such subspaces is

$$\begin{bmatrix} 3 \\ 1 \end{bmatrix}_q = \frac{q^3 - 1}{q - 1} = q^2 + q + 1.$$

Let $\mathcal{B}$ be the collection of two-dimensional subspaces of $\mathbb{F}_q^3$. Their number is also $q^2 + q + 1$. We join a point $p$ of $\mathcal{P}$ to a block $B$ of $\mathcal{B}$ if $p$ is a subspace of $B$. This is a $2 - (q^2 + q + 1, q + 1, 1)$ design. This design is an example of a **projective plane** of order $q$. Notice that for $q = 2$, we have $q^2 + q + 1 = 7$ and our setup corresponds to the Fano plane.

It is unknown if there are any projective planes of order $n$ when $n$ is not a prime power. We will see some necessary conditions for the existence of such object later in this chapter using the Bruck-Ryser-Chowla theorem.

A $t - (v, k, \lambda)$ design consists of a collection $\mathcal{P}$ of $v$ points and a collection $\mathcal{B}$ of blocks (that are $k$-subsets of points) such that for any set of $t$ points, there are exactly $\lambda$ blocks that contain them. A trivial example of a $t$-design consists of all the $k$-subsets of a set of $v$ points.

**Example 9.2.5** If we consider a $v$ element set $\mathcal{P}$ and consider the collection $\mathcal{B}$ of all $k$-element subsets of $X$, we see that any $t$-element set with $0 \le t \le k$ is contained in precisely $\binom{v-t}{k-t}$ blocks. This is a $t - \left( v, k, \binom{v-t}{k-t} \right) -$ design.

A less trivial example is below.

**Example 9.2.6** Consider the following incidence structure, where the points are the 21 edges of the complete graph $K_7$. Each block consists of 5 edges and is of one of the following types:

- a star $K_{1,5}$,
- a cycle $C_5$, and
- a disjoint union of a $K_3$ and two disjoint edges.

The reader can verify that this is $3 - (21, 5, 3)$ design.

## 9.3   Incidence Matrices

A convenient way of encoding the information in a block design $(X, B)$ is by the use of its **incidence matrix**. This is a $v \times b$ matrix $N$ whose rows are labelled by the points of the design and whose columns correspond to the blocks, where

$$N(p, B) = \begin{cases} 1, & \text{if } p \in B \\ 0, & \text{otherwise.} \end{cases}$$

**Theorem 9.3.1** *If $N$ is the incidence matrix of a $2 - (v, k, \lambda)$ block design $(\mathcal{P}, \mathcal{B})$, then*

$$N N^T = \lambda J_v + (r - \lambda) I_v.$$

**Proof** We immediately see that the entries in every column add up to $k$. This means that the diagonal entries of $N N^T$ are $r$. For $p \ne q \in \mathcal{P}$, note that

$$(N N^T)(p, q) = \sum_{B \in \mathcal{B}} N(p, B) N(q, B),$$

which is the same as the number of blocks containing both $p$ and $q$. Since $(\mathcal{P}, \mathcal{B})$ is a $2 - (v, k, \lambda)$ design, this number equals $\lambda$. This proves our assertion. ∎

This theorem allows us to obtain further necessary conditions for the existence of block designs. Indeed, we can compute the determinant of $NN^T$ as follows. The matrices $(r - \lambda)J_v$ and $\lambda I_v$ commute and since their eigenvalues are known (see Example 4.2.2), we deduce that the eigenvalues of $NN^T$ are $rk$ with multiplicity one and $r - \lambda$ with multiplicity $v - 1$. Hence,

$$\det(NN^T) = rk(r - \lambda)^{v-1}.$$

The following result is commonly known Fisher's inequality as it was proved by the English statistician Ronald A. Fisher (1890–1962) in 1940.

**Corollary 9.3.2** (Fisher's inequality) *In any $2 - (v, k, \lambda)$ design with $v > k$, we must have $b \geq v$. That is, there must be at least as many blocks as points.*

*Proof* Theorem 9.2.1 and $v > k$ imply that $r > \lambda$. From the calculation above of the determinant of $NN^T$, we see that the matrix $NN^T$ is non-singular, and thus has rank $v$. Recall that for any two matrices $A$ and $B$ for which $AB$ is defined, the row space of $AB$ is contained in the row space of $A$ (see Exercise 9.6.7). Thus, $\mathrm{rank}(AB) \leq \mathrm{rank}(A)$. In our situation, $v = \mathrm{rank}(NN^T) \leq \mathrm{rank}(N) \leq \min(v, b) \leq b$ as claimed. $\blacksquare$

Designs in which $b = v$ are called **symmetric designs**.

**Corollary 9.3.3** *In a symmetric $2 - (v, k, \lambda)$ design with $v$ even, $k - \lambda$ must be a perfect square.*

*Proof* If $b = v$, we must have that $r = k$ from Theorem 9.2.1. Also, the incidence matrix $N$ is a square matrix and

$$\det(N)^2 = \det(NN^T) = rk(r - \lambda)^{v-1} = r^2(r - \lambda)^{v-1}.$$

Since $N$ is an integer matrix, its determinant is an integer. Hence, the right-hand side is a perfect square and so $(r - \lambda)^{v-1} = (k - \lambda)^{v-1}$ must also be a perfect square. As $v - 1$ is odd, this forces $k - \lambda$ to be a perfect square. $\blacksquare$

**Example 9.3.4** There is no $2 - (22, 7, 2)$ design because $7 - 2$ is not a perfect square.

One may wonder what happens when $v$ is odd. We will prove later the following important theorem in the theory of designs. This was proved in 1949 by the American mathematicians Richard Hubert Bruck (1914–1991) and Herbert John Ryser (1923–1985) and extended in 1950 by the Indian mathematician Sarvadaman Chowla (1907–1995) and Ryser. The result below is known as the Bruck-Ryser-Chowla theorem or the BRC theorem.

**Theorem 9.3.5** (Bruck-Ryser-Chowla) *If $(\mathcal{P}, \mathcal{B})$ is a symmetric $2 - (v, k, \lambda)$ design and $v$ is odd, then the equation*

$$(k - \lambda)x^2 + (-1)^{(v-1)/2}\lambda y^2 = z^2$$

*has a non-zero solution in integers.*

We will give a proof of this result in the next section. We give some applications of it below.

**Example 9.3.6** As an application of this theorem, consider the existence of a $2 - (29, 8, 2)$ design. That is, can we arrange 29 objects into 29 blocks, with each object occurring in 8 blocks and any 2 objects occur in precisely 2 blocks. The theorem implies that if such a design exists, then we can solve the Diophantine equation

$$6x^2 + 2y^2 = z^2$$

with $(x, y, z) \neq (0, 0, 0)$. We may assume that $\gcd(x, y, z) = 1$, as otherwise, we can cancel the common factor. From the equation, we see that 2 divides the left-hand side and hence must divide the right-hand side. So write $z = 2z_1$. We get

$$3x^2 + y^2 = 2z_1^2$$

has a non-trivial solution. If we reduce this mod 3, we get

$$2z_1^2 \equiv y^2 (\mod 3).$$

If $z_1$ is coprime to 3, we deduce that 2 is a square mod 3, which is not the case. Thus, 3 divides $z_1$, so write $z_1 = 3z_2$ to deduce that

$$3x^2 + y^2 = 18z_2^2$$

has a non-trivial solution. But now, 3 divides $y$ and $9|3x^2$ implies $3|x$, contrary to the coprimality assumption at the outset. Hence, there is no such design.

We now prove the only non-existence theorem known in the theory of projective planes.

**Theorem 9.3.7** *If there is a projective plane of order n and $n \equiv 1$ or 2 (mod 4), then n is the sum of two integer squares.*

**Proof** As observed earlier, we are asking for the existence of a $2 - (n^2 + n + 1, n + 1, 1)$ design. Notice that $v = n(n + 1) + 1$ is odd. Applying the Bruck-Ryser-Chowla theorem, we deduce that the Diophantine equation

$$nx^2 + (-1)^{n(n+1)/2}y^2 = z^2$$

has a non-trivial integral solution. If $n \equiv 1$ (mod 4), then $n(n + 1)/2$ is odd, so the theorem says that we can solve

$$nx^2 = z^2 + y^2$$

in non-zero integers. The same implication occurs when $n \equiv 2 \pmod 4$. Thus, $n$ is the sum of two rational squares. To complete the proof, we need to show that $n$ is in fact the sum of two integral squares. Now we need to use one more fact from number theory. Recall that an odd prime number $p$ can be written as a sum of two squares if and only if $p \equiv 1 \pmod 4$. From this, one can deduce that the numbers that can be expressed as a sum of two integer squares are precisely the numbers whose unique factorization into distinct prime powers does not admit a prime $\equiv 3 \pmod 4$ to an odd power. Thus, if $n$ cannot be written as a sum of two squares, then there is a prime $p \equiv 3 \pmod 4$ an odd power $p^{2a+1}$ (say) of which divides $n$ exactly. Reducing the equation mod $p^{2a+1}$, we get

$$y^2 + z^2 \equiv 0 \pmod{p^{2a+1}}.$$

If $y, z$ are coprime to $p$, this is already a contradiction for it says that $-1$ is a perfect square mod $p$. If $y$ and $z$ are not coprime to $p$, only an even power of $p$ can divide each of them and hence both of them and after cancelling it, we still get a contradiction. This completes the proof. ∎

**Example 9.3.8** We can apply this result to show that there is no projective plane of order 6. Indeed, if there is, by the theorem, 6 can be written as a sum of two integral squares, which is clearly not the case. Thus, there is no $2 - (43, 7, 1)$ design. In particular, there is no way to arrange 43 objects into 43 blocks such that each block contains 7 objects and any two objects occurring together in precisely one block.

For a long time, the first unresolved case was $n = 10$. The above theorem does not exclude this possibility, as 10 can be written as $1 + 9$. In 1989, Clement Lam, Larry Thiel, and S. Swiercz from Concordia University, Canada, using the Cray 1 computer showed that there is no projective plane of order 10. However, we still have no conceptual proof of this fact. It is generally believed that projective planes can only exist when $n$ is a prime power, but this has not yet been proved.

## 9.4  Bruck-Ryser-Chowla Theorem

The proof of Theorem 9.3.5 requires the use of Lagrange's four squares theorem.

**Theorem 9.4.1** *Every natural number can be written as a sum of four squares of integer numbers.*

**Proof** We prove it in four steps. As the identity

$$(|z|^2 + |w|^2)(|u|^2 + |v|^2) = |uz - \overline{w}v|^2 + |wu + \overline{z}v|^2$$

is easy to verify directly for all complex numbers $u, v, w, z$, we deduce from it, by putting $z = x_1 + ix_2$, $w = x_3 + ix_4$, $u = y_1 + iy_2$, $v = y_3 + iy_4$ that

$$(x_1^2 + x_2^2 + x_3^2 + x_4^2)(y_1^2 + y_2^2 + y_3^2 + y_4^2) = z_1^2 + z_2^2 + z_3^2 + z_4^2,$$

where

$$z_1 = x_1 y_1 + x_2 y_2 + x_3 y_3 + x_4 y_4$$

$$z_2 = x_1 y_2 - x_2 y_1 + x_3 y_4 - x_4 y_3$$

$$z_3 = x_1 y_3 - x_2 y_4 - x_3 y_1 + x_4 y_2$$

$$z_4 = x_1 y_4 + x_2 y_3 - x_3 y_2 - x_4 y_1.$$

This means that if $a$ can be written as a sum of four integral squares and $b$ can be written as a sum of four integral squares, so can $ab$ and we have an explicit recipe for determining these squares if we know the ones for $a$ and $b$, respectively. As every number is a product of prime numbers, it therefore suffices to prove Lagrange's theorem for prime numbers.

The next step is to see that for any odd prime $p$, we can solve the congruence

$$x^2 + y^2 + 1 \equiv 0 \quad (\text{mod } p).$$

To see this, we consider the set of squares $x^2$ (mod $p$) for $x \in \mathbb{Z}_p$, which has size $1 + (p-1)/2 = (p+1)/2$. The same is true of the set of elements of the form $-1 - y^2$ for $y \in \mathbb{Z}_p$. Hence, there is a common element to these two sets, and this gives a solution to the congruence. Since the integers in the interval $[-(p-1)/2, (p-1)/2]$ forms a complete set of residue classes mod $p$, we may choose $|x| < p/2$ and $|y| < p/2$, we deduce that there are integers $x$ and $y$ so that $x^2 + y^2 + 1 = mp$ with $m < p$.

The third step is to consider the smallest natural number $m$ such that $mp$ can be written as a sum of four squares. By the previous paragraph, the set is non-empty. If $m_0$ be the smallest such integer, then $m_0 < p$. If $m_0 = 1$, we are done so, let us suppose that $1 < m_0 < p$. Hence, we can write

$$m_0 p = x_1^2 + x_2^2 + x_3^2 + x_4^2.$$

If $m_0$ were even, then either all the $x_i$'s are even or all of them are odd, or precisely two of them, say, $x_1, x_2$ (without loss of generality) are even. In any of the cases, $x_1 - x_2, x_1 + x_2, x_3 - x_4, x_3 + x_4$ are even, and we have

$$(m_0/2)p = \left(\frac{x_1 - x_2}{2}\right)^2 + \left(\frac{x_1 + x_2}{2}\right)^2 + \left(\frac{x_3 - x_4}{2}\right)^2 + \left(\frac{x_3 + x_4}{2}\right)^2.$$

Thus, $(m_0/2)p$ can be written as a sum of four squares, and this is a contradiction to the minimality of $m_0$. Hence, $m_0$ is odd.

The final step is choosing $y_1, y_2, y_3, y_4$ so that $y_i \equiv x_i$ (mod $m_0$) with $|y_i| \le (m_0 - 1)/2$. We have that

$$m_0 m_1 = y_1^2 + y_2^2 + y_3^2 + y_4^2,$$

with $m_1 < m_0$. By step 1, we see that $(m_0 p)(m_0 m_1)$ can be written as a sum of four squares:

$$z_1^2 + z_2^2 + z_3^2 + z_4^2,$$

with the $z_i$'s being given explicitly in terms of $x_i$'s and the $y_i$'s. From this explicit description, we see directly that $z_i \equiv 0(\mod m_0)$. Thus, we may divide out by $m_0^2$ and deduce that $m_1 p$ can be written as a sum of four squares. But this contradicts the minimality of $m_0$ as $m_1 < m_0$. Hence $m_0 = 1$. This completes the proof of Lagrange's theorem.                                                                                ∎

Now we will sketch the proof of the Bruck-Ryser-Chowla theorem.

***Proof*** Suppose that we have a symmetric $(v, k, \lambda)$ design with $v$ odd. Let $n = k - \lambda$ and suppose that $v \equiv 3(\mod 4)$. We want to show that $nx^2 = z^2 + \lambda y^2$ has a non-trivial integral solution. It suffices to show that this has a non-trivial rational solution, since we can always clear denominators.

By Lagrange's theorem, we may write $n = a^2 + b^2 + c^2 + d^2$. If we let $H$ be the $4 \times 4$ matrix:

$$\begin{bmatrix} -a & b & c & d \\ b & a & d & -c \\ c & -d & a & b \\ d & c & -b & a \end{bmatrix},$$

then $HH^T = H^T H = nI_4$.

Let $N$ be the incidence matrix of the symmetric block design. This is a $v \times v$ matrix. Consider the $(v + 1) \times (v + 1)$ matrix $B$ obtained by adding a 1 in the $(v + 1, v + 1)$th position and zeros everywhere else in the last row and last column. Then,

$$BB^T = \begin{bmatrix} N^T N & 0 \\ 0 & 1 \end{bmatrix}.$$

As $4|(v + 1)$, we may create the $(v + 1) \times (v + 1)$ matrix $K$ which has $(v + 1)/4$ diagonal blocks of the matrix $H$. Then, $K^T K = K K^T = nI_{v+1}$. Consider the quadratic form

$$x^T B^T Bx = k(x_1^2 + \cdots + x_v^2) + x_{v+1}^2 + \lambda \sum_{1 \le i \ne j \le v} x_i x_j$$

$$= \lambda(x_1 + \cdots + x_v)^2 + x_{v+1}^2 + n(x_1^2 + \cdots + x_v^2).$$

If we put $z = Bx$, then this is

$$\sum_{i=1}^{v+1} z_i^2.$$

Consider another change of coordinates $z = Ky$. Then

$$z^T z = y^T K^T K y$$

which is

$$\sum_{i=1}^{v+1} z_i^2 = n \sum_{i=1}^{v+1} y_i^2.$$

Thus, $x = (B^{-1}K)y$ so that

$$n(y_1^2 + \cdots + y_{v+1}^2) = \lambda(x_1 + \cdots + x_v)^2 + x_{v+1}^2 + n(x_1^2 + \cdots + x_v^2).$$

The idea now is to choose the $x_i$ and $y_i$ suitably to obtain the statement of the theorem. As the matrix $B^{-1}K$ is a rational matrix, we may write

$$x_i = \sum_{i=1}^{v+1} a_i y_i$$

with $a_i$ rational. If $a_1 \neq 1$, choose $x_1 = y_1$; otherwise, choose $x_1 = -y_1$. In either case, $x_1^2 = y_1^2$ and $y_1$ is a rational linear combination of $y_2, \ldots, y_{v+1}$. Thus, $x_2$ is a rational linear combination of $y_2, \ldots, y_{v+1}$:

$$x_2 = \sum_{i=2}^{v+1} b_i y_i,$$

with $b_i$ rational. If $b_2 \neq 1$, choose $x_2 = y_2$; otherwise, choose $x_2 = -y_2$. In either case, $x_2^2 = y_2^2$ and $y_2$ is now a rational linear combination of $y_3, \ldots, y_{v+1}$. We continue in this way for each $i \leq v$, so that $x_i^2 = y_i^2$ for each $i \leq v$ and $y_v$ is a rational multiple of $y_{v+1}$ and $x_{v+1}$ is a rational multiple of $y_{v+1}$. Put $y_{v+1} = 1$. Then, $x_{v+1}$ and $y_v$ are uniquely determined rational numbers and working backwards, so are all the $x_i$'s and the $y_i$'s. Since $x_i^2 = y_i^2$ for $1 \leq i \leq v$, we get that

$$n = ny_{v+1}^2 = \lambda(x_1 + \cdots + x_v)^2 + x_{v+1}^2$$

has a solution in rational numbers. Moreover, the solution is non-trivial since $x_{v+1}$ and $y_{v+1}$ are non-zero. This completes the proof in this case.

The case $v \equiv 1(\mod 4)$ is similar. The essential change in the above proof is that we use the matrix $A$ instead of the matrix $B$ and replace $K$ by the $v \times v$ matrix obtained by putting $H$ on the diagonal and adding a 1 in the $(v, v)$th position and zeros elsewhere in the last row and column. The proof proceeds as before, and we leave it as an exercise to the reader. ∎

## 9.5   Codes and Designs

The fundamental paper *A mathematical theory of communications* from 1948 of the American mathematician Claude Shannon (1916–2001) is considered to be the starting point of coding theory. Around the same time, the American mathematician Richard Wesley Hamming (1915–1998) and the Swiss physicist Marcel J.E. Golay (1902–1989) also significantly contributed to the beginning of this subject.

Let $\mathbb{F}$ be a finite set/alphabet and $n$ be a natural number. **The Hamming distance** $d(u, v)$ between two words $u, v \in \mathbb{F}^n$ is the number of positions where $u$ and $v$ differ. The following result summarizes the basic properties of the Hamming distance.

**Proposition 9.5.1** *Let $\mathbb{F}$ be a finite set and $n$ be a natural number. The Hamming distance has the following properties:*

*(1) (reflexive): $d(u, u) = 0, \forall u \in \mathbb{F}^n$,*
*(2) (symmetric): $d(u, v) = d(v, u), \forall u, v \in \mathbb{F}^n$, and*
*(3) (triangle inequality): $d(u, v) \leq d(u, w) + d(w, v), \forall u, v, w \in \mathbb{F}^n$.*

**Proof** The first two parts are obvious. For the third part, note that $d(u, v)$ equals the smallest number of positions we have to change in $u$ in order to transform it into $v$. ∎

A **code** $\mathcal{C}$ is a subset of $\mathbb{F}^n$. **The length** of $\mathcal{C}$ is $n$. The vectors in the code $\mathcal{C}$ are called **codewords**. **The minimum distance** $d(\mathcal{C})$ is defined as the minimum of $d(u, v)$, taken over all $u \neq v \in \mathcal{C}$. The code $\mathcal{C}$ is an $(n, M, t)$-code over $\mathbb{F}$ if $|\mathcal{C}| = M$ and $d(\mathcal{C}) = t$. The code $\mathcal{C}$ is **binary** if $|\mathbb{F}| = 2$.

**Example 9.5.2** The following collection of vectors

$$c_1 = 00001$$
$$c_2 = 00110$$
$$c_3 = 11111$$
$$c_4 = 11000,$$

forms a $(5, 4, 3)$-code over $\mathbb{F}_2$.

A code $\mathcal{C}$ is said to be $e$-**error correcting** if $d(\mathcal{C}) \geq 2e + 1$. The reason for this definition is given by the following theorem.

**Theorem 9.5.3** *A code $\mathcal{C}$ is $e$-error correcting if and only if the Hamming balls $B_e(c) := \{v : d(v, c) \leq e\}$ are pairwise disjoint for all $c \in \mathcal{C}$.*

**Proof** If $B_e(c_1)$ and $B_e(c_2)$ are not disjoint for two distinct codewords $c_1 \neq c_2 \in \mathcal{C}$, then let $v$ be a common element of these two Hamming balls. By the triangle inequality, we get that

$$d(c_1, c_2) \leq d(c_1, v) + d(v, c_2) \leq 2e,$$

contradiction with $d(c_1, c_2) \geq 2e + 1$.

Conversely, if all the Hamming balls are pairwise disjoint, and $C$ is not $e$-error correcting, then there are two codewords $c_1, c_2$ such that $d(c_1, c_2) = f \leq 2e$. This means that $c_1$ and $c_2$ agree in $f$ positions. Now change the coordinates of $c_1$ in $\lfloor f/2 \rfloor$ of these positions to those of $c_2$ and call this changed vector $b$. We have that

$$d(c_1, b) = \lfloor f/2 \rfloor \leq e, \quad d(c_2, b) = \lceil f/2 \rceil \leq e$$

so that $b \in B_e(c_1) \cap B_e(c_2)$. This completes the proof.                                  ∎

The application of these ideas in communication networks is as follows. If $C$ is an $e$-error-correcting code, then the codewords of $C$ are used to send signals over a *noisy channel*. If a codeword $c \in C$ is sent and at most $e$ errors are made in the transmission, and we receive the word $c' \in \mathbb{F}^n$, then $d(c, c') \leq e$. Thus $c'$ lies in the Hamming sphere $B_e(c)$. By Theorem 9.5.3, $c$ is the unique codeword satisfying this inequality, and we can deduce that $c$ was sent instead of $c'$.

Another application of the previous result is the following constraint on the size of a code in terms of its length and minimum distance. This is usually called the Hamming bound or the sphere-packing bound.

**Theorem 9.5.4** *If $C$ is a $(n, M, d)$-code with $d \geq 2e + 1$, then*

$$M \leq \frac{q^n}{\sum_{j=0}^{e} \binom{n}{j}(q-1)^j}. \tag{9.1}$$

***Proof*** For any word $x \in \mathbb{F}_q^n$, we have that

$$|B_e(x)| = \sum_{j=0}^{e} \binom{n}{j}(q-1)^j.$$

From the previous result, we deduce that

$$q^n = |\mathbb{F}_q^n| \geq |\cup_{u \in C} B_e(c)| = \sum_{c \in C} |B_e(c)|$$

$$= M \cdot \sum_{j=0}^{e} \binom{n}{j}(q-1)^j,$$

which proves our assertion.                                                                        ∎

A code is called **perfect** if it attains equality in the previous bound.

**Example 9.5.5** Let $\mathbb{F}$ be a finite set and $n$ be a natural number. **The repetition code** of length $n$ over $\mathbb{F}$ consists of the codewords of the form $(a, \ldots, a)$, $a \in \mathbb{F}$. If $|F| = q$, then this is a $(n, q, n)$-code. If $n$ is odd and $q = 2$, this code is perfect.

Another less trivial example of a perfect code is obtained by using the Fano plane.

**Example 9.5.6**  Using Fig. 9.1, we consider the characteristic vectors $c_1, \ldots, c_7$ of the 7 triples in the Fano plane and their complements $\overline{c}_1, \ldots, \overline{c}_7$:

$$c_1 = 1101000, \ \overline{c}_1 = 0010111$$
$$c_2 = 1010001, \ \overline{c}_2 = 0101110$$
$$c_3 = 1000110, \ \overline{c}_3 = 0111001$$
$$c_4 = 0110100, \ \overline{c}_4 = 1001011$$
$$c_5 = 0100011, \ \overline{c}_5 = 1011100$$
$$c_6 = 0011010, \ \overline{c}_6 = 1100101$$
$$c_7 = 0001101, \ \overline{c}_7 = 1110010$$

and all zero vector $c_0 = 0000000$ and the all one vector $\overline{c}_0 = 1111111$. The collection of these 16 codewords is a $(7, 16, 3)$-code, which is perfect.

A code is called **linear** if it is a subspace of $\mathbb{F}_q^n$, for some prime power $q$ and natural number $n$. The **weight** of a vector $v$, denoted $wt(v)$, is the number of non-zero coordinates of $v$.

**Proposition 9.5.7**  *If $C$ is a linear code, then $d(C) = \min_{u \in C^*} w(u)$, where $C^*$ denotes set of non-zero codewords of $C$.*

**Proof**  If $x, y \in \mathbb{F}_q^n$, then $w(x - y) = d(x, y)$. The result follows since $C$ is a vector subspace of $\mathbb{F}_q^n$.  ∎

A linear code $C$ is called a $[n, k, t]$-linear code over $\mathbb{F}_q$ if it has length $n$, its dimension as a subspace of $\mathbb{F}_q^n$ is $k$ and its minimum distance is $t$. A $k \times n$ matrix $G$ whose rows form a basis for $C$ is called a **generator matrix** of $C$. **The dual code** $C^\perp$ of $C$ consists of all the words of $\mathbb{F}_q^n$ that are orthogonal to the rows of $G$ (and to all the vectors in $C$). A $(n - k) \times n$ matrix $H$ whose rows form a basis for the dual code $C^\perp$ is called a **parity-check matrix** for $C$.

**Proposition 9.5.8**  *Let $C$ be a linear code with a parity-check matrix $H$. The minimum distance of $C$ equals the maximum $t$, such that any $t - 1$ columns of $H$ are linearly independent (and some $t$ columns of $H$ are linearly dependent).*

**Proof**  Let $h_1, \ldots, h_n$ be the columns of $H$. The codewords of $C$ are the vectors $\begin{bmatrix} x_1 \ldots x_n \end{bmatrix}^T$ such that $x_1 h_1 + \cdots + x_n h_n = 0$. The result now follows from using Proposition 9.5.7.  ∎

**Example 9.5.9**  Let $r \geq 1$. Define the $r \times (2^r - 1)$ matrix $H_{r,2}$ whose columns are the non-zero vectors of $\mathbb{F}_2^r$. When $r = 3$, we have that

$$H_3 = \begin{bmatrix} 0 & 0 & 0 & 1 & 1 & 1 & 1 \\ 0 & 1 & 1 & 0 & 0 & 1 & 1 \\ 1 & 0 & 1 & 0 & 1 & 0 & 1 \end{bmatrix}.$$

Consider the binary code of length $2^r - 1$ whose parity-check matrix is $H_{r,2}$. This is called **the binary Hamming code**. The reader can use the previous proposition to show that this is a $[2^r - 1, 2^r - 1 - r, 3]$-linear code and Theorem 9.5.4 to show that it is a perfect code.

There is a generalization of the previous construction.

**Example 9.5.10** Let $q$ be a prime power and $r$ a natural number. Denote $m = \frac{q^r-1}{q-1}$. Consider now the $r \times m$ matrix whose columns are spanning vectors for the $m$ one-dimensional subspaces of $\mathbb{F}_q^r$. **The $q$-ary Hamming code** is the linear code whose parity-check matrix is $H_{r,q}$. As above, one can show that this code is a $[m, m - r, 3]$-linear code that is perfect.

**Example 9.5.11** Consider the code whose codewords are the rows of the incidence matrix of a symmetric $(v, k, \lambda)$-design. Each codeword has weight $k$ and for any two distinct codewords, there are exactly $\lambda$ positions where they have entry one. If $R_1$ and $R_2$ are distinct rows, then $d(R_1, R_2) = 2(k - \lambda)$ Thus, the rows of a symmetric $(v, k, \lambda)$-design give us a $(k - \lambda - 1)$-error-correcting code.

**Example 9.5.12** In 1971, the Mars Mariner spacecraft used the rows of a $(31, 15, 7)$-design as codewords to send back photographs of Mars back to Earth. This code corrects 7 errors. In later space missions, more sophisticated codes called Reed-Solomon codes have been used, and these codes are capable of correcting a larger number of errors. They are based on the following simple idea. Given a codeword $(a_0, a_1, \ldots, a_{m-1})$, construct a polynomial $f(x) = a_0 + a_1 x + \cdots + a_{m-1} x^{m-1}$. Fix a primitive root $g$ of $\mathbb{F}_q$. Instead of trying to send the codeword, the spacecraft transmits the sequence

$$f(0), f(g), f(g^2), \ldots, f(g^N),$$

where $N > m$. Since a polynomial of degree $m$ is determined by $m + 1$ values, this is sufficient information to retrieve the original codeword $(a_0, \ldots, a_{m-1})$. One can prove that this method gives rise to a $(q + m)/2$-error-correcting code.

## 9.6   Exercises

**Exercise 9.6.1** If $n \geq k$ are natural numbers, prove that

$$(q^k - 1) \begin{bmatrix} n \\ k \end{bmatrix}_q = (q^n - 1) \begin{bmatrix} n - 1 \\ k - 1 \end{bmatrix}_q.$$

**Exercise 9.6.2** If $n \geq k$ are natural numbers, prove that

$$\begin{bmatrix} n + 1 \\ k \end{bmatrix}_q = \begin{bmatrix} n \\ k - 1 \end{bmatrix}_q + \begin{bmatrix} n \\ k \end{bmatrix}_q + (q^n - 1) \begin{bmatrix} n - 1 \\ k - 1 \end{bmatrix}_q.$$

**Exercise 9.6.3** Let $f_q(n)$ be the number of subspaces of $\mathbb{F}_q^n$. Show that

$$f_q(n+1) = 2f_q(n) + (q^n - 1)f_q(n-1).$$

**Exercise 9.6.4** Let $L$ be the lattice of subspaces of $\mathbb{F}_q^n$ partially ordered by inclusion. If $W$ is a subspace of dimension $k$, show that the Möbius function for this lattice is given by

$$\mu(0, W) = (-1)^k q^{\binom{k}{2}}.$$

**Exercise 9.6.5** A group of 16 students decides to sign up for three field trips each. Each trip accommodates precisely 6 students. The students would like to sign up in such a way that any two of them would be together on precisely one of the trips. Is such an arrangement possible?

**Exercise 9.6.6** Consider the following incidence structure: the points are the edges of the complete graph $K_6$ and the blocks are all the sets of three edges that form a perfect matching or a triangle in $K_6$. Show that this is a Steiner triple system on 15 points.

**Exercise 9.6.7** Show that the number of blocks in a $t - (v, k, \lambda)$ design equals $\lambda \binom{v}{t} / \binom{k}{t}$.

**Exercise 9.6.8** For any natural number $n \geq 1$, consider the incidence structure, where the points are the non-zero vectors of $\mathbb{F}_2^n$ and the blocks are the triples $\{x, y, z\}$ with $x + y + z = 0$. Show that this a $2 - (2^n - 1, 3, 1)$ design.

**Exercise 9.6.9** If $A$ is a $m \times n$ matrix and $B$ is a $n \times p$ matrix over a field $\mathbb{F}$, show that

$$\text{rank}(AB) \leq \min(\text{rank}(A), \text{rank}(B)).$$

**Exercise 9.6.10** In a symmetric $2 - (v, k, \lambda)$ design with incidence matrix $N$, show that

$$\frac{1}{k - \lambda}\left(N + \sqrt{\frac{\lambda}{k}}J_v\right)$$

is the inverse of

$$N^T - \sqrt{\frac{\lambda}{v}}J_v.$$

Deduce that

$$N^T N = \lambda J_v + (r - \lambda)I_v.$$

Use this equation to prove that in any symmetric design, every pair of blocks has precisely $\lambda$ elements in common.

**Exercise 9.6.11** Show that there is no projective plane of order 14.

**Exercise 9.6.12** If $p \equiv 3 \pmod 4$ is a prime, show that there is no $2 - (v, p + 1, 1)$ design with $v \equiv 3 \pmod 4$.

**Exercise 9.6.13** Prove that the Hamming codes in Example 9.5.10 are perfect.

**Exercise 9.6.14** If $C$ is a code in $\mathbb{F}_q^n$ with distance $d(C) = d$, then

$$|C| \le q^{n-d+1}.$$

**Exercise 9.6.15** Let $n$ be a natural number. Assume that we have two pairwise $n \times n$ orthogonal Latin squares $L_1$ and $L_2$ over an alphabet $\mathbb{F}$ of order $n$. Construct the code $\mathcal{C}$ whose codewords are the $n^2$ words of length 4 of the form:

$$(i, j, L_1(i, j), L_2(i, j)),$$

for any $1 \le i, j \le n$. Show that this code has minimum distance 3 and attains equality in the previous bound.

**Exercise 9.6.16** Label the points of the Fano plane by the elements of $\mathbb{Z}_7$ such that each block of the Fano plane has the form $\{x, x + 1, x + 3\}$ for $x \in \mathbb{Z}_7$. For $n \in \{3, 4\}$, find a subset $S_n$ of $\mathbb{Z}_{n^2+n+1}$ such that the elements of $\mathbb{Z}_{n^2+n+1}$ as points and the $n^2 + n + 1$ lines $S_n + x = \{s + x : s \in S_n\}$ form a projective plane of order $n$.

**Exercise 9.6.17** Show that if a $2 - (v, 3, 1)$ design exists, then prove that $v \equiv 1, 3 \pmod 6$.

**Exercise 9.6.18** Consider the design whose point set is $\mathbb{Z}_n \times \mathbb{Z}_3$. The blocks are the triples $\{(x, 0), (x, 1), (x, 2)\}$ for $x \in \mathbb{Z}_n$ and $\{(x, i), (y, i), (\frac{x+y}{2}, i + 1)\}$ for $x \ne y \in \mathbb{Z}_n$ and $i \in \mathbb{Z}_3$. Show that this is a $2 - (6t + 3, 3, 1)$-design or Steiner triple system of order $6v + 3$.

**Exercise 9.6.19** If there is a $t - (v, k, 1)$ design with $v > k + t$, prove that $v \ge (t + 1)(k - t + 1)$.

**Exercise 9.6.20** Show that there are at most two disjoint Steiner triple systems on a set of 7 points.

## Reference

E. Brown, K. Mellinger, Kirkman's schoolgirl wearing hats and walking through fields of numbers. Math. Mag. **82**, 3–15 (2009)

# Chapter 10
# Planar Graphs

## 10.1 Euler's Formula

**A curve** in the plane is the image of a continuous map from $[0, 1]$ to $\mathbb{R}^2$. A graph is said to be **embedded** in the plane if it can be drawn on the plane so that its vertices correspond to distinct points of the plane, its edges are represented by curves connecting their corresponding endpoints and no two edges/curves intersect except only at their endpoints. Such a graph is called a **planar graph**. A planar embedding $Y$ of a planar graph $X$ is a graph isomorphic to $X$, which we call a **plane graph**, and we refer to the vertices of $Y$ as points and to its edges as lines.

**A region** in the plane is an open set $W$ such that for any $x \neq y \in W$, $W$ contains a curve from $x$ to $y$. Given a connected plane graph $Y$, **a face** of $Y$ is a maximal region of the plane which does not contain any points of $Y$. A connected plane graph has also one unbounded face, called the **outer face**. **The length** is the length of the closed walk around the edges bounding the face. For the plane graph in Fig. 10.1, there are three faces of length 4. **The dual graph** $X^*$ of a plane graph $X$ is the graph whose vertices are the faces of $X$. For any edge of $X$ *separating* two (not necessarily distinct) faces, we draw an edge; the vertices of $X^*$ corresponding to those faces of $X$.

In Fig. 10.2, we have a plane graph on four vertices whose vertices are filled dots. This graph has two faces, one of length 3 (the triangle) and the outerface of length 5. Its dual is a multigraph with two vertices drawn as small circles and has one loop (at the vertex corresponding to the outer face) and three other multiple edges (one for each edge of the triangle).

The basic relation for planar graphs is the following theorem due to the Swiss mathematician Leonhard Euler (1707–1783).

**Theorem 10.1.1** (Euler, 1758) *If $X$ is a connected plane graph with $v$ vertices, $e$ edges and $f$ faces, then*

$$v - e + f = 2. \tag{10.1.1}$$

© Hindustan Book Agency 2022
S. M. Cioabă and M. R. Murty, *A First Course in Graph Theory and Combinatorics*,
Texts and Readings in Mathematics 55, https://doi.org/10.1007/978-981-19-0957-3_10

**Fig. 10.1** The graph $K_{2,3}$
and a planar embedding

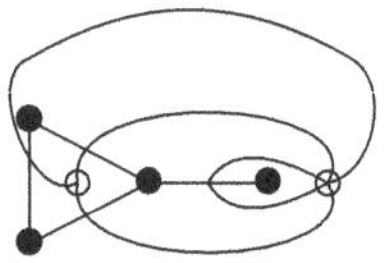

**Fig. 10.2** A plane graph and
its dual

*Proof* There are many proofs of this result. We give one using induction on the number of vertices. If $v = 1$, then $X$ is a collection of loops at a single vertex. If $e = 0$, then $f = 1$ and the formula is true in this case. Each added loop cuts the face into two faces and so increases the face count by one. Hence, the formula holds in case $v = 1$.

Let $X$ be a connected plane graph with $v \geq 2$ vertices. Consider an edge $e_0$ which is not a loop. The contraction of $X$ by $e_0$ gives a plane graph $X/e_0$. Contraction does not reduce the number of faces, so $X/e_0$ has $v - 1$ vertices, $e - 1$ edges, and $f$ faces. Since $X/e_0$ has fewer number of vertices, we can apply the induction hypothesis to get that

$$2 = (v - 1) - (e - 1) + f = 2 = v - e + f,$$

which is what we want to prove.                                                                    ∎

If $X$ is not connected, then Euler's formula fails. If $X$ is a planar graph with $c$ connected components, then $v - e + f = c + 1$. This is easily seen by adding $c - 1$ edges or bridges between the $c$ components and then applying Euler's formula to the resulting connected graph. Adding the bridges does not alter the face count. Thus, we get that $v - (e + c - 1) + f = 2$ from which the formula follows.

Euler's formula has many applications. The first consequence is that planar graphs cannot have too many edges.

**Theorem 10.1.2** *If $X$ is a simple planar graph with at least $v \geq 3$ vertices, then $e \leq 3v - 6$. If $X$ is triangle-free, then $e \leq 2v - 4$.*

*Proof* It suffices to prove the result for connected graphs. Every face of $X$ has at least three edges, and each edge appears in two faces. Thus, $3f \leq 2e$ and using Euler's formula (10.1.1) gives us $2 = v - e + f \leq v - e + \frac{2e}{3}$ which implies the desired inequality $e \leq 3v - 6$. If $X$ is triangle-free, then each face contributes at least four edges. Since each edge appears in two faces, we get $2e \geq 4f$. Putting this back into Euler's formula gives the second inequality.                                                                    ∎

This result can be generalized to get an upper bound for the number of edges of a planar graph of a given girth (see Exercise 10.6.3).

**Corollary 10.1.3** *The graphs $K_5$ and $K_{3,3}$ are non-planar (Fig. 10.3).*

**Fig. 10.3** Two non-planar
graphs $K_{3,3}$ and $K_5$

**Fig. 10.4** A subdivision of
$K_{3,3}$

**Fig. 10.5** The Petersen
graph can be contracted to
$K_5$

*Proof* If the graph $K_5$ were planar, then applying the theorem gives $10 \leq 15 - 6 = 9$, a contradiction. For $K_{3,3}$, we get that $9 \leq 18 - 6 = 12$ which does not give a contradiction if we use the first inequality. However, the bipartite graph has no triangles and so, by the second inequality, we get $9 \leq 8$, which is a contradiction. ∎

Given a graph $X$, a **subdivision** of $X$ is obtained from $X$ replacing its edges with paths. It is not too hard to see that a graph is planar if any subdivision of it is planar (Fig. 10.4).

Kazimierz Kuratowski (1896–1980) was a Polish mathematician who made important contributions to topology and combinatorics. A famous theorem of Kuratowski from 1930 gives a characterization of planar graphs.

**Theorem 10.1.4** (Kuratowski, 1930) *A graph is non-planar if and only it contains a subgraph that is a subdivision of $K_5$ or $K_{3,3}$.*

Another formulation is of this result is that a graph is not planar if and only if it can be edge-contracted to a $K_5$ or a $K_{3,3}$. The Petersen graph (shown below) can be contracted to the complete graph $K_5$ by collapsing the edges connecting the *outer* $C_5$ to the *inner* $C_5$ and therefore, it is not planar. One can avoid the use of a *big hammer* (Kuratowski's theorem) to solve this problem and instead use a generalization of Theorem 10.1.2 to show that the Petersen graph is not planar (see Exercise 10.6.4). The reader can also check that the Petersen graph contains a subdivision of $K_{3,3}$ so it fails being planar in a spectacular way (Fig. 10.5).

**Theorem 10.1.5** *Every simple planar graph $X$ contains a vertex of degree at most five.*

*Proof* By contradiction, if every vertex has degree at least six, then $6v \leq 2e$ which gives that $3v \leq e \leq 3v - 6$, contradiction with Theorem 10.1.2. ∎

As a warm-up for the next section, we can prove the six-colour theorem.

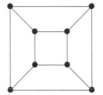 

**Fig. 10.6** The tetrahedron, the cube, and the octahedron

**Theorem 10.1.6** *Every planar graph can be properly coloured using six colours.*

**Proof** We proceed by induction on the number of points of the plane graph. The result is true for any such graph with 5 or fewer points. Let $v \geq 6$ be a natural number. Suppose that all plane graphs with fewer $v - 1$ points or less are 6-colourable. Let $X$ be a plane graph with $v$ points. By Theorem 10.1.5, $X$ contains a vertex of degree 5 or less. By induction, $X - v$ is 6-colourable. As $v$ has degree 5 or less, we can colour it with one of the six colours not used on any of its adjacent vertices. ∎

## 10.2   The Platonic Solids

The ancient Greeks were fond of geometry and believed that some regular solids were associated with the elements: air, earth, fire, and water. For example, air was represented by the octahedron, earth was associated with the *three*-dimensional cube, fire with the tetrahedron, and water with the icosahedron. We leave it to the reader to find out what Plato and Aristotle wrote about the dodecahedron.

A **Platonic solid** is a convex polyhedron whose faces are regular polygons having the same number of edges, such that each vertex is contained in the same number of faces. A Platonic solid can be embedded in the plane as a regular plane graph where all the faces have the same length.

**Proposition 10.2.1** *Let $X$ be a $k$-regular plane graph whose faces have length $\ell$. If $k \geq 3$ and $\ell \geq 3$, then*

- $(k, \ell) = (3, 3)$ *(tetrahedron),*
- $(k, \ell) = (3, 4)$ *(cube),*
- $(k, \ell) = (4, 3)$ *(octahedron),*
- $(k, \ell) = (3, 5)$ *(icosahedron),*
- $(k, \ell) = (5, 3)$ *(dodecahedron).*

**Proof** By double counting, we get that $2e = f\ell = kv$. Euler's formula implies that $2 = v - e + f = e(2/k - 1 + 2/\ell)$, and therefore, $2/k + 2/\ell > 1$. This is equivalent to $(k - 2)(\ell - 2) < 4$. Since $k, \ell \geq 3$, the only solutions are the ones claimed above. ∎

**Fig. 10.7**  The icosahedron

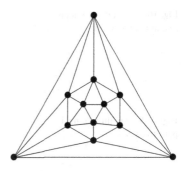

The cube and the octahedron are dual to each other while the tetrahedron is self-dual. The icosahedron is below, while the dodecahedron is in Fig. 11.2. They are dual to each other (Figs. 10.6 and 10.7).

## 10.3  The Five Colour Theorem

The Four Colour Theorem has a colourful history! It states that any planar graph can be coloured using four colours. Since the complete graph $K_4$ is planar and has chromatic number four, we see that four colours are necessary. To prove that this is sufficient for any planar graph is much more difficult. The Four Colour Conjecture was first formulated by Francis Guthrie on October 23, 1852. Guthrie was a student at University College London, where he studied under Augustus de Morgan (1806–1871). When Francis Guthrie asked de Morgan, he did not know how to prove it and wrote to Sir William Rowan Hamilton (1805–1865) in Dublin to ask if he knew how to prove it. It seems that Guthrie graduated and then studied law. After practicing as a barrister, he went to South Africa in 1861 as a professor of mathematics. After a few mathematical papers, he switched to the field of botany.

In the meanwhile, Augustus de Morgan circulated Guthrie's question to many mathematicians. Arthur Cayley, who learned of the question from de Morgan in 1878, posed it as a formal unsolved problem to the London Mathematical Society on 13 June 1878. On 17 July 1879, Alfred Kempe (1849–1922), a London barrister and amateur mathematician, announced that he had a proof. Kempe had studied under Cayley, and at Cayley's suggestion, submitted his paper to the journal *American Journal of Mathematics* in 1879. Apparently, Kempe received great acclaim for his work. He was elected Fellow of the Royal Society and served as its treasurer for many years. In 1912, he was knighted. The error in his *proof* was discovered in 1890 by Percy John Heawood (1861–1955), a lecturer in Durham, England. In his paper, Heawood showed how to salvage the proof and prove that every map is 5-colourable.

**Theorem 10.3.1**  (Heawood) *Any planar graph is 5-colourable.*

**Fig. 10.8**  A vertex of degree
five in a plane graph

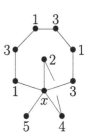

**Fig. 10.9**  An alternating 1,3
path from $p$ to $r$

***Proof***  We will use induction on the number of vertices. The result is true for any
planar graph with five or fewer vertices. Let $v \geq 6$ and consider a planar graph $X$
with $v$ vertices. By Theorem 10.1.5, there is a vertex $x$ of degree five or less. By
induction, $X - x$ can be properly coloured using 5 colours. If the degree of $x$ is four
or less, then colour $x$ with a colour not used for any of its adjacent vertices.

Hence, we may assume that $x$ has degree five. Let us label the neighbours of $x$ as
$p, q, r, s, t$ as in Fig. 10.8. If there is one colour missing on these neighbours, then we
can colour $x$ with that colour and obtain a proper colouring of $X$. Assume that this is
not the case and that $p, q, r, s, t$ are coloured with colours 1, 2, 3, 4, 5, respectively.

Denote by $X_{i,j}$ the subgraph of $X - x$ induced by those vertices whose colours are
$i$ and $j$. Consider $X_{1,3}$. Both $p$ and $r$ belong to $X_{1,3}$. If they lie in distinct components
of $X_{1,3}$ then, we may interchange the colours 1 and 3 in the component containing $r$.
Note that this is still a proper colouring, but now $r$ has colour 1. We can now colour $x$
using colour 3 and have a proper colouring of $X$. If $p$ and $q$ lie in the same connected
component of $X_{1,3}$, then there is a path of vertices with alternating colours 1 and 3
from $p$ to $r$.

Now consider $X_{2,4}$. Both $q$ and $s$ belong to this subgraph. Again, if $q$ and $s$ lie in
distinct connected components of $X_{2,4}$, we may interchange colours 2 and 4 in the
component of $q$ and obtain another proper colouring of $X$ where both $q$ and $s$ have
colour 4. We can then colour $x$ with colour 2 and have a proper colouring of $X$. If $q$
and $s$ lie in the same component of $X_{2,4}$, then there is a path of alternating colours
from $q$ to $s$. However, this path must cross the path from $p$ to $r$ described earlier (see
Fig. 10.9) which contradicts the fact that $X$ is a planar graph. Thus, this possibility
cannot arise and $X$ can be properly coloured with five colours.  ∎

In 1977, the American mathematician Kenneth Appel (1932–2013) and the German mathematician Wolfgang Haken succeeded in proving the Four Colour theorem
Appel and Haken (1977), Appel et al. (2022). Their work was assisted by 1200 h
of computer work at that time. Later on, their proof was simplified, but still with

Fig. 10.10 The graph $K_5$ on
a torus

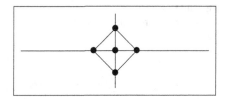

computer assistance by Neil Robertson, Daniel Sanders, Paul Seymour, and Robin
Thomas (1997).

## 10.4 Colouring Graphs on Surfaces

The graph $K_5$ is not a planar graph. However, it can be drawn on a torus without any
crossings (Fig. 10.10).

A celebrated theorem of Augustus Möbius from 1870 is that any compact (orientable) surface is homeomorphic to a sphere with $g$ handles. The **genus** of the
surface is denoted $g$. A torus, for example, has genus one since it is homeomorphic
to a surface with one handle.

One can show that any graph $X$ can be embedded in some compact orientable
surface. The minimal genus of the surface for which this can be done is called the
**genus of the graph**. For example, the genus of $K_5$ is one and the reader is invited to
show that the genus of $K_{3,3}$ is also 1.

For graphs embedded on a surface of genus $g$, Euler's formula generalizes as
follows. A **face** is defined as a maximal region cut out by the graph which contains
no vertex of the graph in its interior. We state the following result without proof.

**Theorem 10.4.1** (Euler's formula) *For a connected graph of genus g with v vertices,
e edges and f faces,*

$$v - e + f = 2 - 2g.$$

Using Euler's formula, we can prove as before that any simple graph of genus
$g$ has at most $3(v - 2 + 2g)$ edges. This is the analogue that a planar graph has at
most $3(v - 2)$ edges. As before, the hand-shaking lemma gives $2e \leq 6(v - 2 + 2g)$
so that there has to be at least one vertex of degree at most $\frac{6(v-2+2g)}{v}$. This is the
analogue of the result for planar graph which says there is at least one vertex of
degree at most five. We get the following result.

**Theorem 10.4.2** (Heawood, 1890) *The chromatic number of a graph X of genus*
$g > 0$ *is at most* $\left\lfloor \frac{7+\sqrt{1+48g}}{2} \right\rfloor$.

Notice that if $g = 0$ were allowed in the formula, then we deduce the Four Colour
Theorem.

**Fig. 10.11** A polygon
whose corners are lattice
points

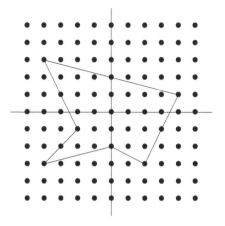

***Proof*** Let $c = \frac{7+\sqrt{1+48g}}{2}$. If $X$ has at most $c$ vertices, we are done. So suppose that $v > c$. If we can show that every simple graph of genus $g$ has a vertex of degree at most $c - 1$, then we can use an induction argument as before to complete the proof. Notice that $c^2 - 7c + (12 - 12g) = 0$ so that $c - 1 = 6 + \frac{12(g-1)}{c}$. Thus, from the remark before the statement of the theorem, we have that $X$ has a vertex of degree at most $6 + \frac{12(g-1)}{v} \leq 6 + \frac{12(g-1)}{c} = c - 1$ as desired.                                    ∎

Notice that $g \geq 1$ has a key role in the inequalities at the end of the proof. For a long time, it was an outstanding problem to determine the genus of the complete graph. The **complete graph conjecture**, proved in 1968 by the German mathematician Gerhard Ringel (1919–2008) and the Indian-born, American mathematician John Wheaton Theodore (Ted) Youngs, states that the genus of $K_n$ is $\left\lceil \frac{(n-3)(n-4)}{12} \right\rceil$.

## 10.5   Pick's Theorem

A point in the Cartesian plane is called **a lattice point** if its coordinates are integers. Consider the polygon in Fig. 10.11 whose corners are lattice points. What is its area?

A beautiful theorem from 1899 due to the Austrian mathematician Georg Alexander Pick (1859–1942) gives the answer by counting the number of lattice points on its boundary and in its interior.

**Theorem 10.5.1** (Pick, 1899) *Let $P$ be a polygon whose corners are lattice points. If $i$ is the number of its interior points and $b$ is the number of its border points, then the area of $P$ equals $i + b/2 - 1$.*

The key to proving this result is the particular case of it.

**Lemma 10.5.2** *A triangle whose corners are lattice points is called* **primitive** *if it has no lattice points in its interior nor on the border except the corners (equivalently, $i = 3$ and $b = 0$). The area of any primitive triangle is $1/2$.*

**Fig. 10.12** A primitive
triangle

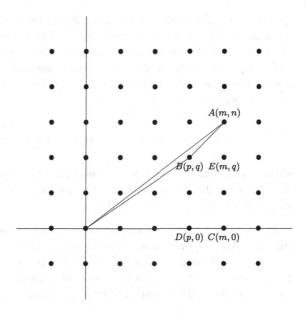

*Proof* First, note that if $(m, n)$ is a non-zero lattice point, then the number of lattice points on the closed segment $(0, 0)$ to $(m, n)$ equals $\gcd(m, n) + 1$. Each point $(r, s)$ on this segment must satisfy the equation $rn = sm$ which implies that $r \cdot \frac{n}{\gcd(m,n)} = s \cdot \frac{m}{\gcd(m,n)}$. Since $\frac{m}{\gcd(m,n)}$ and $\frac{n}{\gcd(m,n)}$ are coprime, we deduce that $\frac{m}{\gcd(m,n)}$ divides $r$. Since $0 \le r \le m$, we get that $r = \frac{m}{\gcd(m,n)} \cdot t$ for an integer $t$ with $0 \le t \le \gcd(m, n)$.

Second, if we consider a triangle whose corners are the lattice points $(0, 0)$, $(m, 0)$ and $(m, n)$, where $m$ and $n$ are natural numbers, then the number of boundary lattice points is $m + n + \gcd(m, n)$ and the number of interior lattice points equals $\frac{(m-1)(n-1)-\gcd(m,n)+1}{2}$. When $\gcd(m, n) = 1$, these numbers are $m + n + 1$ and $\frac{(m-1)(n-1)}{2}$, respectively.

By translation and rotation, we may assume that the primitive triangle is as in Fig. 10.12. If $O$ denotes the origin, we can calculate the area of the triangle $OAB$ as

$$\text{area}(OAC) - \text{area}(OBD) - \text{area}(BECD) - \text{area}(ABE)$$
$$= \frac{mn}{2} - \frac{pq}{2} - (m - p)q - \frac{(m - p)(n - q)}{2},$$

which is $\frac{np-mq}{2}$. It remains to show that $np - mq = 1$. Because the triangle $OAB$ is primitive, we must have that $\gcd(m, n) = \gcd(p, q) = \gcd(m - p, n - q) = 1$. By counting the number of interior lattice points in the above triangles and rectangle, we leave to the reader to complete the proof that $np - mq = 1$. ∎

*Proof* We now prove Pick's theorem. The proof uses Euler's formula for planar graphs (Theorem 10.1.1). Triangulate the polygon into primitive triangles using the lattice points from its interior and from the boundary. One can regard the triangulated polygon as a plane graph. If $v$ is the number of its vertices, $e$ is the number of its edges and $f$ the number of its faces, Euler's formula implies that $v - e + f = 1$ (since we ignore the infinite face). Hence $f = e - v + 1$. Now, the number of its vertices equals $i + b$. Each interior edge is contained in two triangles, and each edge on the boundary is contained in one triangle. It follows that $2e = 3f + b$. Combining these equations, we get that

$$f = e - v + 1 = \frac{3f + b}{2} - (i + b) + 1 = \frac{3f}{2} - i - b/2 + 1,$$

which implies that $f = 2i + b - 2$. Since each face is a primitive triangle whose area is $1/2$ and these triangles decompose the polygon, we deduce that the area of the polygon must be $\frac{2i+b-2}{2} = i + b/2 - 1$.                                   ∎

Another proof of Pick's formula can be given by triangulating the polygon into primitive triangles and summing up the angles in those triangles. For our example in Fig. 10.11, we have that $i = 23$ and $b = 10$ and therefore, the area must be $23 + 10/2 - 1 = 27$. The book Michael (2009) contains other applications of Pick's formula, and we highly recommend it for the curious reader.

## 10.6   Exercises

**Exercise 10.6.1** Show that the graph obtained by removing one edge from the complete graph $K_5$ is planar.

**Exercise 10.6.2** Show that the graph obtained by removing one edge from the complete bipartite graph $K_{3,3}$ is planar.

**Exercise 10.6.3** **The girth** of a graph $X$ is the shortest length of one of its shortest cycle, if such a cycle exists or $\infty$ otherwise. Use Euler's formula to show that if $X$ is a connected planar graph with finite girth $g$, $v$ vertices and $e$ edges, then

$$e \leq \frac{g(v - 2)}{g - 2}.$$

**Exercise 10.6.4** Determine the girth of the Petersen graph (see Fig. 10.5) and use the previous question to deduce that it is not a planar graph.

**Exercise 10.6.5** Determine all natural numbers $r$ and $s$ such that $K_{r,s}$ is a planar graph.

**Exercise 10.6.6** Let $X$ be a graph with chromatic number $\chi(X) > 3$. Show that the genus $g(X)$ of a graph $X$ satisfies the inequality

$$g(X) \geq \frac{1}{12}\left(\chi(X)^2 - 7\chi(X) + 12\right).$$

Deduce that for $n \geq 5$, the genus of the complete graph $K_n$ is at least

$$\left\lceil \frac{(n-3)(n-4)}{12} \right\rceil.$$

**Exercise 10.6.7** Let $G$ be the graph obtained from $K_{4,4}$ by deleting a perfect matching. Is $G$ planar ? If yes, give a drawing. If no, give a proof.

**Exercise 10.6.8** Let $S$ be a set of $n$ points in the plane such that the distance between any two of them is at least 1. Show that there are at most $3n - 6$ pairs $x$, $y$ such that the distance between $x$ and $y$ is 1.

**Exercise 10.6.9** **The crossing number** of a graph $X$ is the minimum number of crossings in a drawing of $X$ in the plane. Determine the crossing number of $K_5$ and $K_{3,3}$.

**Exercise 10.6.10** Let $X$ be a graph with $v$ vertices and $e$ edges. If $k$ is the maximum number of edges in a planar subgraph of $X$, show that the crossing number of $X$ is at least $e - k$. Prove that the crossing number of $X$ is at least $e - 3v + 6$. If $X$ has no triangles, then the crossing number is at least $e - 2v + 4$.

**Exercise 10.6.11** Show that the crossing number of $K_6$ is 3.

**Exercise 10.6.12** Determine the crossing number of the Petersen graph.

**Exercise 10.6.13** A planar graph $X$ is **outerplanar** if it has a drawing with every vertex on the boundary of the unbounded face. Show that any cycle is outerplanar. Show that $K_4$ is planar, but not outerplanar.

**Exercise 10.6.14** Prove that $K_{2,3}$ is planar, but not outerplanar.

**Exercise 10.6.15** Prove that every outerplanar graph has a vertex of degree at most 2.

**Exercise 10.6.16** Any outerplanar graph is 3-colourable.

**Exercise 10.6.17** An art gallery is represented by a polygon with $n$ sides. Show that it is possible to place $\lfloor \frac{n}{3} \rfloor$ guards such that every point interior to the polygon is visible to some guard. Construct a polygon that can be guarded by precisely $\lfloor \frac{n}{3} \rfloor$ guards.

**Exercise 10.6.18** Show that every planar graph decomposes into two bipartite graphs.

**Exercise 10.6.19** For any $n \geq 4$, construct a planar graph with $n$ vertices and chromatic number 4.

**Exercise 10.6.20** Use Pick's formula to find the number of ways to make change for $N$ dollars using quarters (25 cents), dimes (10 cents), and nickels (5 cents)?

# References

K. Appel, W. Haken, Every planar map is four colorable. I. Discharging. Ill. J. Math. **21**, 429–490 (1977)

K. Appel, W. Haken, J. Koch, Every planar map is four colorable. II. Reducibility. Ill. J. Math. **21**, 491–567

T.S. Michael, *How to Guard an Art Gallery and Other Discrete Mathematical Adventures* (Johns Hopkins University Press, Baltimore, 2009)

N. Robertson, D. Sanders, P. Seymour, R. Thomas, The four-colour theorem. J. Comb. Theory, Ser. B **70**, 2–44 (1997)

# Chapter 11
# Edges and Cycles

## 11.1 Edge Colourings

In the previous chapters, we have been considering vertex colourings. Now we will consider edge colourings of a graph. We will say that two edges are **incident** if they have a common vertex. We would like to properly colour the edges, in the sense that no two incident edges receive the same colour. Given a graph $X$, **the edge chromatic number** or **the chromatic index** $\chi'(X)$ is defined as the minimum number of colours needed to properly colour the edges.

The question of edge colourings occurs in many contexts. Suppose that in a school we have $n$ teachers $T_1, \ldots, T_n$ to teach $m$ classes $C_1, \ldots, C_m$ and teacher $T_i$ must teach class $C_j$ for $p_{ij}$ class periods. Is it possible to schedule these classes in such a way that a minimum number of time slots are used? To study this question, we would construct a bipartite graph whose vertices consist of vertices $T_i$ and the classes $C_j$. We will join $T_i$ to $C_j$ by $p_{ij}$ multiple edges. An edge colouring of this graph corresponds to a timetable for the school. It will turn out that the edge chromatic number for this graph is $p$ which is the maximum of the values of $p_{ij}$.

If $X$ is a simple graph whose vertex maximum degree is $\Delta$, then we will need at least $\Delta$ colours to properly edge colour $X$. The following remarkable theorem proved by the Ukrainian mathematician Vadim Vizing (1937–2017) in 1964 states that $\chi'(X)$ can only take one of two possible values.

**Theorem 11.1.1** (Vizing, 1964) *If $X$ is a simple graph, then*

$$\Delta \leq \chi' \leq \Delta + 1.$$

Thus, the edge chromatic number of a graph is either $\Delta$ or $\Delta + 1$ and at present time, there is no efficient algorithm to determine the exact value of $\chi'$ for a given graph $X$.

However, in some cases, it is possible to determine which one of these occurs as the edge chromatic number.

© Hindustan Book Agency 2022
S. M. Cioabă and M. R. Murty, *A First Course in Graph Theory and Combinatorics*,
Texts and Readings in Mathematics 55, https://doi.org/10.1007/978-981-19-0957-3_11

**Fig. 11.1** Five disjoint perfect matchings in $K_6$

**Theorem 11.1.2** *If $X$ is a $\Delta$-regular graph with an odd number of vertices, then* $\chi' = \Delta + 1$.

***Proof*** Assume the number of vertices is $2k + 1$. In any proper edge colouring, the edges coloured with the same colour form a matching. Hence, the number of edges in any such colour class is at most $k$. Therefore, $\chi' \geq \frac{(2k+1)\Delta}{2k} > \Delta$. By Vizing's theorem, $\chi'$ must equal $\Delta + 1$.                                                          ∎

We can apply this result to determine the edge chromatic number of the complete graph.

**Corollary 11.1.3** *The edge chromatic number of $K_n$ is $n - 1$ if $n$ is even and $n$ if $n$ is odd.*

***Proof*** First, suppose $n$ is odd. As $K_n$ is $(n - 1)$-regular, we can apply the previous theorem, to deduce that we will need at least $n$ colours.

When $n$ is even, we can draw the vertices of it on the plane as follows. Represent all but one of the vertices of $K_n$ as the points of a regular $(n - 1)$-gon in the plane. The remaining vertex is represented by the centre of the polygon. Construct a perfect matching by joining the centre vertex with one of the other points. Pair up the remaining $n - 2$ vertices in edges/segments perpendicular to the edge/segment above. This creates one perfect matching in $K_{2n}$. Rotating this perfect matching clockwise by $2\pi/(n - 1)$ creates another perfect matching that is disjoint from the first one. Repeating this procedure creates $n - 1$ pairwise disjoint perfect matchings. See Fig. 11.1 for an illustration of $n = 6$.

Each perfect matching is a colour class and this shows that $\chi'(K_n) \leq n - 1$. Therefore, $\chi'(K_n) = n - 1$ when $n$ is even.                                                          ∎

Thus, $K_n$ with $n$ odd requires one more colour than the maximal degree. Another class of graphs with this property is the cycle graph $C_n$ whose edge chromatic number is easily seen to be 3 if $n$ is odd and 2 if $n$ is even. The edge colouring problem can be transformed into a vertex colouring problem in the following way. Given a graph $X$, we consider the **line graph** of $X$, denoted by $\mathcal{L}(X)$, whose vertices represent the edges of $X$. Two vertices of $\mathcal{L}(X)$ are adjacent if the corresponding edges in $X$ meet in a vertex. The edge chromatic number of $X$ is the chromatic number of its line graph.

We can determine the edge chromatic number of bipartite graphs by a celebrated theorem of Dénes König (1884–1944). The method of proof resembles the five colour theorem.

**Theorem 11.1.4** (König) *If $X$ is a bipartite graph with maximal degree $\Delta$, then $\chi'(X) = \Delta$.*

***Proof*** We use induction on the number of edges of $X$. If $X$ has no edges or one edge, then the statement is true. Assume that $X$ is a bipartite graph with two or more edges. Let $U$ and $V$ be the partite sets of $X$. Remove an edge $e$ that joins $u \in U$ and $v \in V$. The resulting graph $X'$ is still bipartite. The highest vertex degree in $X'$ is still $\Delta$ or less. By induction, we can edge colour $X'$ using $\Delta$ colours or less. If we used at most $\Delta - 1$ colours, then we can use the remaining colour for the edge $e$ and that would finish the proof. Otherwise, assume we used $\Delta$ colours to properly colour the edges of $X'$. If one of the colours is missing from both $v$ and $u$, then we can use that colour to colour $e$ and we would be done again. So assume otherwise. Since the degree of $u$ in $X'$ is $\Delta - 1$ or less, there is a colour $c$ missing at $u$ which is used at an edge incident with $v$. Similarly, a colour $c'$ is missing at $v$ and $c'$ is used at one edge incident with $u$. Consider the subgraph of $X'$ induced by the edges coloured with either $c$ or $c'$. In particular, consider the component which contains $u$. This component is a path whose edges have alternating colours ($c'$ and $c$). Because $X$ is a bipartite graph, this path cannot contain $v$. Now we can perform a colour reversal (switch the colours $c$ and $c'$) in this component and still have a proper edge colouring of $X'$ with colour $c$ used now at $u$ instead of $c'$. As $c'$ is missing from $v$, we can use that colour to colour the edge $e$. This completes the proof. ∎

An application of Theorem 11.1.4 is to the problem of scheduling. Suppose that members of a hiring committee are to interview a number of candidates for a job. If each member interviews at most $m$ candidates and each candidate is interviewed by at most $n$ members individually, then Theorem 11.1.4 tells us that $\max(m, n)$ time slots are needed for this purpose. The precise timetabling of the candidates corresponds to matchings, which we can effectively determine using the Hungarian algorithm.

**Corollary 11.1.5** *For any natural numbers $r$ and $s$, $\chi'(K_{r,s}) = \max(r, s)$.*

A **decomposition** of a graph is a list of subgraphs such that each edge appears in exactly one subgraph in the list. An edge colouring of the graph gives rise to a decomposition of the edges into colour classes which are matchings of the graph. A $k$-**factor** of a graph is a spanning subgraph which is a $k$-regular graph. Thus, a 1-factor is a matching. If we edge colour a $k$-regular graph using only $k$ colours, then the colour classes give us a decomposition of the graph into 1-factors. Julius Petersen (1839–1910) was a Danish mathematician who in 1898 introduced the graph we call now the Petersen graph (see Fig. 10.5) as an example of a bridgeless cubic graph with edge chromatic number four (see Exercise 11.6.3).

A long-standing open problem in this area is **the total colouring conjecture** due to Vadim Vizing and, independently, the Iranian mathematician Mehdi Behzad from the 1960s. **The total chromatic number** $\chi''(X)$ of a simple graph $X$ is the minimum number of colours used in a colouring of the vertices and the edges of $X$ such that no adjacent vertices have the same colour, no incident edges have the same colour and no edge and its endpoints have the same colour. By Vizing's theorem, we can deduce that $\chi'' \geq \Delta + 1$, where $\Delta$ is the maximum degree of $X$.

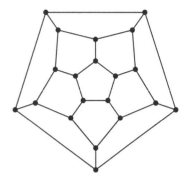

**Fig. 11.2** The dodecahedron graph

**Conjecture 11.1.6** (Behzad-Vizing) *If X is a simple graph with maximum degree* $\Delta$, *then* $\chi'' \leq \Delta + 2$.

## 11.2  Hamiltonian Cycles

A **Hamiltonian cycle** in a graph is a cycle that contains every vertex of the graph. This is reminiscent of the Eulerian circuit discussed in Chap. 1. However, no simple criterion is known that characterizes graphs that contain a Hamiltonian cycle. In a practical context, this arises as to the famous Travelling Salesman Problem. The question is to visit each of the cities in a circuit exactly once and minimizing the cost of such a tour. No efficient algorithm is known for finding such a tour, and it represents one of the major unsolved problems in graph theory.

Despite being studied first by Thomas Kirkman (1806–1895), Hamilton cycles are named for Sir William Rowan Hamilton (1805–1865). In 1856, Kirkman asked if given the graph of a polyhedron, does there exists a cycle passing through every vertex? In 1857, Hamilton invented the Icosian game, which is the problem of finding a Hamilton cycle in a **dodecahedron** (seen in the Fig. 11.2). The game was actually sold as a pegboard with holes at the nodes of the dodecahedron. In 1859, Hamilton sold the game to a London game dealer for 25 pounds, who marketed it under the name *Around the World*.

There is also the famous problem of the *knight's tour*, which asks if the knight on the chessboard can visit each of the squares exactly once and return to the starting point. This problem was first solved by Leonhard Euler (1707–1783) in 1759.

There are many theorems which provide sufficient conditions for the existence of a Hamiltonian cycle. We give one such theorem, due to the Norwegian mathematician Oystein Ore (1899–1968), below.

**Theorem 11.2.1** (Ore) *Let X be a simple graph with $v$ vertices. Suppose that $v \geq 3$ and that* $\deg(x) + \deg(y) \geq v$ *for any pair of non-adjacent vertices x and y. Then, X has a Hamiltonian cycle.*

**Fig. 11.3** An illustration of the proof of Theorem 11.2.1

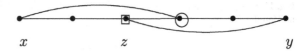

$x$                  $z$                               $y$

*Proof* Suppose that the theorem is false for some graph with $v \geq 3$ vertices. Add as many edges as possible to this graph without producing a Hamiltonian cycle. Call this new graph $X'$. The graph $X'$ cannot be the complete graph and so there is a pair of non-adjacent vertices $x$ and $y$. By the construction of $X'$, any addition of an edge will create a Hamiltonian cycle. In particular, adding the edge $xy$ will create a Hamiltonian cycle. Thus, $X'$ has a path from $x$ to $y$ which visits all the vertices. Of these vertices, put a circle around the ones which are adjacent to $x$ and put a square around the vertices before the circled ones in the Hamiltonian path. There are $\deg(x)$ circled/squared vertices and $v - 1 - \deg(x)$ not squared vertices. Because $\deg(y) \geq v - \deg(x) > v - 1 - \deg(x)$, there is a vertex $z$ that is adjacent to $y$ which has been squared (see Fig. 11.3). Thus, we have a Hamiltonian cycle from $x$ to the vertex *after* $z$ and then moving to $y$ on the Hamiltonian path and then to $z$ and back to $x$. This is a contradiction that $X'$ has a Hamiltonian cycle. ∎

We have the following result due to the Hungarian-English mathematician Gabriel Andrew Dirac (1925–1984).

**Corollary 11.2.2** *If $X$ is a graph with $v$ vertices such that every vertex has degree $\geq v/2$, then $X$ has a Hamiltonian cycle.*

The Petersen graph is a 3-regular graph with edge chromatic number 4 (see Exercise 11.6.3). One can show that any 3-regular graph with a Hamiltonian cycle has edge chromatic number 3 and therefore, we conclude that the Petersen graph is not Hamiltonian (see Exercises 11.6.2 and 11.6.4).

The following is a necessary condition for a graph to have a Hamiltonian cycle.

**Proposition 11.2.3** *If $X$ has a Hamiltonian cycle, then for each $S \subset V(X)$, the graph $X - S$ obtained from $X$ by removing the vertices in $S$, has at most $|S|$ components.*

*Proof* Let $C$ be a Hamiltonian cycle in $X$ and $S$ be a subset of vertices of $X$. When $C$ leaves a component of $X \setminus S$, it must return to $S$. These returns to $S$ must use different vertices of $S$. This proves the theorem. ∎

This result can be useful when trying to show that a graph is not Hamiltonian. The graph below is such an example (Fig. 11.4).

**The toughness** $t(X)$ of a connected graph $X$ is the minimum of $|S|/c(X - S)$, where $S$ is a subset of vertices and $c(X - S)$ is the number of components of $X - S$. The graph $X$ is $t$-tough if $t(X) \geq t$. Hence, the previous result states that a Hamiltonian graph must be 1-tough. The converse is not true, and the graph below is an example of a 1-tough non-Hamiltonian graph.

A long-standing open problem in graph theory is the following conjecture (Fig. 11.5).

**Fig. 11.4**  A graph that has
no Hamiltonian cycle

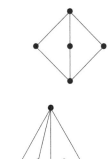

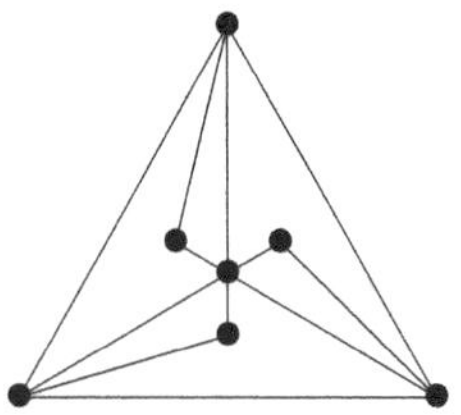

**Fig. 11.5**  A 1-tough
non-Hamiltonian graph

**Conjecture 11.2.4**  *There exists $t_0$ such that if $X$ is $t$-tough with $t \geq t_0$, then $X$ is Hamiltonian.*

A directed complete graph is called a **tournament**, since it can be used to record the outcome of a round-robin tournament (where every contestant is matched against every other contestant). The arrows on the edges would indicate the win-loss record.

**Theorem 11.2.5**  *A tournament always contains a Hamiltonian path.*

**Proof** Consider a path of the longest length. The claim is that this path contains every vertex. Suppose that $(x_1, \ldots, x_m)$ is our path, and say that $x_0$ is not included in it. If $(x_0, x_1)$ is an edge, then we can add this to our path and get a longer path in which all the vertices are distinct. Thus, $(x_1, x_0)$ is in the tournament. If $(x_0, x_2)$ were an arc in our tournament, we can create the longer path $(x_1, x_0, x_2, \ldots, x_m)$, which would give us a contradiction. Repeating this argument, we deduce that $(x_m, x_0)$ is in the tournament. However, this means that we have the path $(x_1, \ldots, x_m, x_0)$ which is longer than our maximum path. This contradiction finishes our proof. ∎

Theorem 11.2.5 has applications to the job-sequencing problem, where we must arrange jobs to be performed in a sequence so as no time is wasted. For instance, suppose $n$ books are to be printed and then bound. There is one printing machine and one binding machine. Let $p_i$ denote the printing time of the $i$th book and $b_i$ the binding time for the $i$th book. For any two books, $i$ and $j$, we know that either $p_i \leq b_j$ or $p_j \leq b_i$. Theorem 11.2.5 tells us that it is possible to specify the order in which the books are printed and then bound so that the binding machine will be kept busy until all the books are bound once the first book is printed. Thus, the total time for completing the task is $p_k + \sum_{i=1}^{n} b_i$ for some $k$. Indeed, we construct a directed complete graph on $n$ vertices; there is a directed edge from $i$ to $j$ if and only if $b_i \geq p_j$. A Hamiltonian path is an ordering of the books satisfying the above condition.

**Fig. 11.6** The Turán graph
$T_{8,3}$

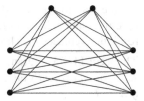

## 11.3 Turán's Theorem

In this section, we describe a famous result in graph theory, called Turán's theorem, which determines the maximum number of edges in a graph of $n$ vertices that does not contain a clique with $r + 1$ vertices. Paul Turán (1910–1976) was a famous Hungarian mathematician with important contributions to number theory, probability, graph theory, and combinatorics. The above-mentioned theorem was proved by Turán while in a concentration camp in 1941 and is considered the starting point of extremal graph theory, an area of research studying how global graph parameters such as number of edges, chromatic number, or girth, say, influence the local structure of the graph.

**Definition 11.3.1** Let $n \geq r \geq 1$ be two natural numbers. The Turán graph $T_{n,r}$ is the complete $r$-partite graph of order $n$ whose partite classes have sizes $\lfloor n/r \rfloor$ or $\lceil n/r \rceil$.

Note that for given $n \geq r \geq 1$, the Turán graph $T_{n,r}$ is well-defined and unique. To see this, let $q = \lfloor n/r \rfloor$. Then by integer division $n = qr + s$, where $0 \leq s \leq r - 1$. The Turán graph $T_{n,r}$ has $s$ partite sets of size $q + 1$ and $r - s$ partite sets of size $q$. It turns out that the Turán graph $T_{n,r}$ has the largest number of edges among all graphs with chromatic number $r$. The starting point of this section is the observation that having more edges in a graph will increase its chromatic number (Fig. 11.6).

**Proposition 11.3.2** *If $X$ is a graph of order $n$ with chromatic number $\chi(X) = r$, then*

$$e(X) \leq e(T_{n,r}).$$

*Equality happens if and only if $X$ is the Turán graph $T_{n,r}$.*

**Proof** Let $X$ be a graph with chromatic number $r$. Then its vertices can be partitioned into $r$ independent sets of sizes $a_1, \ldots, a_r$, say. Clearly, $e(X) \leq e(K_{a_1,\ldots,a_r})$. If two sizes, say $a_j$ and $a_\ell$ differ by at least 2, $a_j \geq a_\ell + 2$, then we move one vertex $x$ from the $j$th partite set to the $\ell$th partite set. The vertex $x$ loses $a_\ell$ neighbours and gains $a_j - 1$ neighbours in this process, while the edges not involving $x$ stay the same. Overall, the new complete $r$-partite graph will have $a_j - 1 - a_\ell + e(K_{a_1,\ldots,a_r}) > e(K_{a_1,\ldots,a_r})$ edges. Hence, the maximum of the number of edges in $K_{a_1,\ldots,a_r}$ will be attained when $|a_j - a_\ell| \leq 1$. It is not hard to see that the only complete $r$-partite graph with this property is the Turán graph $T_{n,r}$. ∎

One way to rephrase the previous result is to state its contrapositive, if $X$ has $n$ vertices and $e(X) > e(T_{n,r})$, then $\chi(X) \geq r + 1$. Turán's theorem is the stronger result that replaces the chromatic number by the clique number.

**Theorem 11.3.3** (Turán, 1941) *Let $X$ be a graph with $n$ vertices that contains no $K_{r+1}$, or equivalently $\omega(X) \leq r$. Then $e(X) \leq e(T_{n,r})$. Equality happens if and only if $X = T_{n,r}$.*

***Proof*** We will prove a slightly stronger result, namely, that for any graph $X$ on $n$ vertices with $\omega(X) \leq r$, there exists a $r$-partite graph $X'$ (so $\chi(X') \leq r$) such that $e(X) \leq e(X')$. If this result is true, then combining it with Proposition 11.3.2 will prove Turán's theorem.

We will use induction on $r$ to prove the stronger result mentioned above. For $r = 1$, if $X$ is a graph not containing $K_2$, then $X$ must be the empty graph on $n$ vertices which is the same as the Turán graph $T_{n,1}$. Hence, the base case is true.

Let $r \geq 2$. Assume that our statement above is true for $r - 1$. Let $X$ be a graph on $n$ vertices with $\omega(X) \leq r$. Let $\Delta$ denote the maximum degree of $X$ and consider a vertex $x$ of $X$ whose degree is $\Delta$. Let $X_1$ denote the subgraph of $X$ induced by the neighbours of $x$. Because $X$ contains no $K_{r+1}$, we deduce that $X_1$ does not contain any $K_r$. By the induction hypothesis, there exists a $(r - 1)$-partite graph $X_1'$ on $\Delta$ vertices such that $e(X_1) \leq e(X_1')$.

Denote by $S$ the subset of $X$ consisting of $x$ and its non-neighbours. Clearly, $|S| = n - \Delta$. We construct the graph $X'$ from $X$ as follows. We keep the edges from $x$ to its neighbours, but we replace the subgraph $X_1$ induced by the neighbours of $x$ by $X_1'$. We also delete any edges of $X$ that may have both endpoints in $S$ and we add all the edges between $S$ and its complement. In the graph $X'$, the subset $S$ is an independent set and since $X_1'$ is a $(r - 1)$-partite graph, this means that $X'$ is a $r$-partite graph. Using $e(X_1) \leq e(X_1')$, we get that

$$e(X) = e(X_1) + e(S, X_1) + e(S) \leq e(X_1) + \sum_{y \in S} d_X(y)$$

$$\leq e(X_1') + \sum_{y \in S} d_X(y) \leq e(X_1') + \Delta |S|$$

$$= e(X_1') + \Delta(n - \Delta)$$

$$= e(X').$$

This completes our proof.                                                                                 ∎

The attentive reader will have noticed that the previous theorem generalizes an earlier result (Exercise 1.6.20). When $r$ divides $n$,

$$e(T_{n,r}) = \binom{r}{2} \left(\frac{n}{r}\right)^2 = \left(1 - \frac{1}{r}\right) \cdot \frac{n^2}{2}.$$

The calculation is a bit more involved when $r$ does not divide $n$, but one can show relatively easily that

$$e(T_{n,r}) \sim \left(1 - \frac{1}{r}\right)\binom{n}{2},$$

when $r$ is fixed and $n$ is large.

## 11.4  Ramsey Theory

The pigeonhole principle is a simple, yet important tool in combinatorics and mathematics. It says that if we have $n$ pigeonholes and $n + 1$ objects are placed in these pigeonholes, then there is at least one pigeonhole with two objects. Though this sounds like a simple principle, its mode of application can be very ingenious at times, leading to striking results.

The pigeonhole principle is essentially an averaging argument.

**Proposition 11.4.1** *Let $n$ be a natural number and $x_1, \ldots, x_n$ be $n$ integers. There exists $j$ such that $x_j \geq \left\lceil \frac{x_1 + \cdots + x_n}{n} \right\rceil$, and there exists $\ell$ such that $x_\ell \leq \left\lfloor \frac{x_1 + \cdots + x_n}{n} \right\rfloor$.*

**Proof** Take $j$ and $\ell$ such that $x_j = \max\{x_1, \ldots, x_n\}$ and $x_\ell = \min\{x_1, \ldots, x_n\}$. The result follows because $x_j$ and $x_\ell$ are integers. ∎

This proposition implies the pigeonhole principle in the following way.

**Corollary 11.4.2** *Let $k$ and $n$ be two natural numbers. If we place $kn + 1$ objects into $n$ pigeonholes, then one of the pigeonholes contains at least $k + 1$ objects and another contains at most $k$ objects.*

**Proof** Indeed, if $x_i$ denotes the number of objects in the $i$th pigeonhole, then the average of the $x_i$'s is $k + 1/n$ and so there is a $j$ such that $x_j \geq \lceil k + 1/n \rceil = k + 1$ and there is a $\ell$ such that $x_\ell \leq \lfloor k + 1/n \rfloor = k$. ∎

Here are some simple examples of the pigeonhole principle in action. The key to the applications is figuring out the pigeonholes.

**Example 11.4.3** In any simple graph, there are two vertices of the same degree. Suppose there are $n$ vertices. The degrees are elements of the set $\{0, 1, \ldots, n - 1\}$. However, having a vertex of degree $n - 1$ means no vertices of degree 0 and vice versa. In either situation, the set of possible degrees has size $n - 1$ and since we have $n$ vertices, two must have the same degree.

**Example 11.4.4** Let $n$ be a natural number. In any $n + 1$ numbers chosen from the set $\{1, 2, \ldots, 2n\}$, two of them must be consecutive. The pigeonholes are the $n$ sets of the form $\{2k - 1, 2k\}$ for $1 \leq k \leq n$.

**Example 11.4.5** Let $n$ be a natural number. No matter how we choose $n + 1$ numbers from the set $\{1, \ldots, 2n\}$, there are two such that one divides the other. For $1 \leq k \leq n$, the $k$th pigeonhole consists of all numbers $m$ in $\{1, \ldots, 2n\}$ such that $m/(2k - 1)$ is a power of two.

A vast generalization of the pigeonhole principle falls under the name of Ramsey theory. Frank Plumpton Ramsey (1906–1930) was an English mathematician, economist, and philosopher who made important contributions in these areas. The basic idea of Ramsey theory can be summarized by the words of Theodore Samuel Motzkin (1908–1970):

*Complete disorder is impossible.*

The simplest case of Ramsey theory can be illustrated by the following amusing fact.

**Proposition 11.4.6** *In any group of six people, there are either three mutual friends or three mutual strangers.*

**Proof** This can be stated in graph-theoretic terms as follows. Take the complete graph $K_6$ whose vertices represent the six people. Colour an edge red if the two people know each other, and colour it blue, otherwise. We claim that there must be a monochromatic triangle.

To see this, let $x$ be a vertex. There must be three edges of the same colour, say red, incident with $x$. If any one of the edges connecting these three vertices is red, we are done, as we have found a red triangle. Otherwise, we have a blue triangle, and we are done again.                                                                                           ∎

A reformulation of the previous result is that any graph of order 6 contains either a clique of order 3 or an independent set of order 3. The value 6 is the best possible, as the cycle $C_5$ on five vertices has both clique number and independence number equal to 2.

Ramsey's generalization of the previous proposition can be stated as follows.

**Theorem 11.4.7** (Ramsey, 1930) *Let $a, b \geq 2$ be integers. If $n = \binom{a+b-2}{a-1}$, then any edge colouring of $K_n$ with red and blue contains a red $K_a$ or a blue $K_b$.*

**Proof** The proof proceeds by induction on $a + b$. The case $a = 2$ or $b = 2$ is trivial. For the induction step, let $a \geq 3$ and $b \geq 3$ and assume that the statement is true for $a + b - 1$. In particular, if $n_1 = \binom{a+(b-1)-2}{a-1}$, in any red/blue edge colouring of $K_{n_1}$, then there is a red $K_a$ or a blue $K_{b-1}$. Also, if $n_2 = \binom{(a-1)+b-2}{(a-1)-1}$ then for any red/blue edge colouring of $K_{n_2}$, there is a red $K_{a-1}$ or a blue $K_b$.

Let $n = \binom{a+b-2}{a-1}$. Consider an arbitrary red/blue edge colouring of the edges of $K_n$. Consider one vertex $x$ and its incident $n - 1$ edges. By the recurrence for the binomial coefficients, we have that

$$n - 1 = \binom{a+b-2}{a-1} - 1 = \binom{a+b-3}{a-1} + \binom{a+b-3}{a-2} - 1$$
$$= n_1 + n_2 - 1$$
$$> (n_1 - 1) + (n_2 - 1).$$

Thus, among these $n - 1$ edges incident with $x$ must be at least $n_1$ blue edges or $n_2$ red edges. Let us consider the first case. The complete graph $K_{n_1}$ vertices, formed by the $n_1$ vertices incident with the above blue edges, contains a red $K_a$ or a blue $K_{b-1}$. If the former is the case, we are done. If the latter, then together with the vertex $x$, we have the required $K_b$. The second case can be solved similarly, and we leave the details to the reader. ∎

For natural numbers $a$ and $b$, **the Ramsey number** $R(a, b)$ is defined to be the smallest value of $n$ such that any red/blue edge colouring of $K_n$ contains a red $K_a$ or a blue $K_b$. Our previous remarks imply that $R(3, 3) = 6$. Our theorem above shows that in general

$$R(a, b) \leq \binom{a + b - 2}{a - 1}.$$

While the exact values of the Ramsey numbers $R(a, b)$ with $a = 3, b \in \{3, \ldots, 9\}$, or $a = 4, b \in \{4, 5\}$ have been determined, the remaining values of $R(a, b)$ are unknown at this time (see the survey Radziszowski 1994). The precise determination of the rest of these numbers is still a major unsolved problem.

There are several ways in which the previous theorem has been generalized. One way is to extend it to more colours than just two colours.

**Theorem 11.4.8** *Let $r \geq 2$ and $a_1, \ldots, a_r \geq 2$ be some natural numbers. There exists a n such that if $K_n$ is edge coloured using colours $1, \ldots, r$, then there is an $i \in \{1, \ldots, r\}$ and a $K_{a_i}$ whose edges are all coloured with the ith colour.*

*Proof* We use induction on $r$. The case $r = 2$ has been proved in Theorem 11.4.7. Assume now that $r > 2$. Applying the induction hypothesis with $r - 1$ and $a_2, \ldots, a_{r-1}$, there is a $n_0$ such that if for any edge colouring of $K_{n_0}$ using $r - 1$ colours, then there is some $K_{a_i}$ with the $i$th colour. Now let $n = R(a_1, n_0)$. Colour the edges of $K_n$ with colours $1, \ldots, r$. By Theorem 11.4.7, there is either a $K_{a_1}$ with the first colour or a $K_{n_0}$ with colours $2, \ldots, r$. In the first situation, we have a $K_{a_1}$ coloured with colour 1 and we are done. Otherwise, by the induction hypothesis, there is $i \in \{2, \ldots, r\}$ and a $K_{a_i}$ whose edges are all coloured with colour $i$ and we are done as well. ∎

Paul Erdős (1913–1996) was a famous Hungarian mathematician who made many important contributions to mathematics, especially in discrete mathematics. He published over 1500 mathematical papers with over 500 collaborators. Among his many results, Erdős introduced the probabilistic method, which will be used to prove the lower bound in the next result.

**Theorem 11.4.9** (Erdős, 1947) *For $a \geq 3$, the diagonal Ramsey number $R(a, a)$ satisfies the inequalities*

$$\lfloor 2^{a/2} \rfloor < R(a, a) \leq 2^{2a}.$$

*Proof* The second inequality follows from Theorem 11.4.7 because

$$\binom{2a - 2}{a - 1} < 2^{2a-2} < 2^{2a}.$$

For the lower bound, we apply a *probabilistic method* as follows. Let $n = \lfloor 2^{a/2} \rfloor$. Consider a random edge colouring of $K_n$ with red and blue, we colour each pair $\{i, j\}$ with $1 \le i < j \le n$, independently red or blue with equal probability $1/2$. For a given subset of vertices $S$ of order $a$, let $A_S$ denote the event that the subgraph induced by $S$ is monochromatic (all its edges have the same colour). It is not hard to see that $\Pr[A_S] = 2^{1-\binom{a}{2}}$. Note that $\binom{n}{a} \le n^a/a!$. Also, the probability that there is a monochromatic $K_a$ equals

$$
\begin{aligned}
\Pr[\cup_{S:|S|=a} A_S] &\le \sum_{S:|S|=a} \Pr[A_S] = \binom{n}{a} 2^{1-\binom{a}{2}} \\
&\le \frac{n^a}{a!} \cdot 2^{1-\binom{a}{2}} \\
&< 1,
\end{aligned}
$$

where the last inequality follows because of $n = \lfloor 2^{a/2} \rfloor \le 2^{a/2}$. Hence, with positive probability, the complement of the even $\cup_{S:|S|=a} A_S$ happens, which means that there is a colouring of $K_n$ with no monochromatic $K_a$. This implies that $R(a, a) > n$ and finishes our proof. ∎

One of the most important open problems in Ramsey theory is determining the asymptotic behaviour of the diagonal Ramsey number $R(a, a)$ when $a$ is large. Determining whether the limit

$$
\lim_{a \to \infty} \sqrt[a]{R(a, a)}
$$

exists and if so, its value, are important open problems in Ramsey theory.

As can be seen, Ramsey theory is a merging of the induction technique with the pigeonhole principle. Its application in many problems can sometimes be quite subtle. The Russian-German mathematician Issai Schur (1875–1941) made important contributions to group representations, combinatorics and number theory. In 1916, Schur applied Theorem 11.4.8 to show that for any sufficiently large prime $p$, the *Fermat equation* mod $p$:

$$
a^m + b^m \equiv c^m \pmod{p}
$$

has a non-trivial solution in integers provided that $p > f(m)$ for some number $f(m)$. Schur's work is based on the following result.

**Theorem 11.4.10** *For any natural number $m$, there is a natural number $M = M(m)$ such that for any partition of $\{1, 2, \ldots, M\}$ into $m$ sets, the equation $x + y = z$ has a solution in at least one of the sets.*

**Proof** By Ramsey's Theorem 11.4.8, there is a $M$ such that if the edges of $K_M$ are coloured using $m$ colours, then there is a monochromatic triangle. Consider a partition of $\{1, \ldots, M\}$ into $m$ sets. Assign a colour from 1 to $m$ to each block of this partition. For each $1 \le i < j \le M$, colour the edge $\{i, j\}$ of $K_M$ with the colour assigned to the block in which $j - i$ appears. Theorem 11.4.8 implies that there is a monochromatic triangle, meaning that there are elements $i < j < k$ such that $j - i, k - j$ and $k - i$ have the same colour, that is, lie in the same block. Hence, $x = k - j, y = j - i, z = k - i$ lie in the same block and $x + y = z$. This finishes the proof. ∎

**Corollary 11.4.11** *For any natural number $m$, there is a natural number $f(m)$ such that for any prime $p > f(m)$, the equation $a^m + b^m \equiv c^m \pmod{p}$, has a nontrivial solution.*

**Proof** Let $f(m) = M$ as in Theorem 11.4.10. Let $p$ be a prime $p > M$. The multiplicative group $\mathbb{F}_p^*$ of non-zero elements is cyclic, and let $g$ be a generator of it. Hence $\{1, \ldots, p - 1\} = \{g^k : 0 \le k \le p - 2\}$. Colour each element of the form $g^{ms+r}$ by colour $r$, where $0 \le ms + r \le p - 1$ and $0 \le r \le m - 1$. We have a colouring with $m$ colours $0, \ldots, m - 1$ of the set $\{1, \ldots, p - 1\}$ of at least $M$ elements. By the previous theorem, there exists a colour $t$ and three elements $x$, $y$ and $z$ of colour $t$ such that $x + y = z \pmod{p}$. Hence, $x = g^{ms_1+t}$, $y = g^{ms_2+t}$ and $z = g^{ms_3+t}$. Plugging these values into $x + y = z$, we get that $(g_1^s)^m + (g_2^s)^m = (g_3^s)^m \pmod{p}$. Taking $a = g_1^s, b = g_2^s$ and $c = g_3^s$, we get that $a, b, c$ are all non-zero since they lie in $\mathbb{F}_p^*$ and $a^m + b^m \equiv c^m \pmod{p}$. This completes the proof. ∎

## 11.5 Graham-Pollak Theorem

While working at the Bell Labs in the 1970s, the American mathematician Ron Graham (1935–2020) and the Austrian-American mathematician Henry Pollak studied the following addressing problem.

**Problem 11.5.1** Given a graph $X = (V, E)$, find an addressing of the vertex set of $X$, namely a function $\ell : V \to \{0, 1, *\}^k$ for some natural number $k$, such that for any two vertices $x, y \in V$, the distance in $X$ between $x$ and $y$ equals the number of positions $j$ where $\{\ell(x)_j, \ell(y)_j\} = \{0, 1\}$.

The origins of this problem are in communication networks, where one would like to label/address the vertices of a graph/network with words of the same length over some alphabet such that the minimum distance in the graph between any two vertices can be determined from the addresses of those vertices. One can try finding such labelling over the alphabet $\{0, 1\}$, but a quick inspection of small graphs such as the complete graph $K_3$ shows that such a task is impossible. The interesting fact is that by adding just one element to the alphabet, namely, the symbol $*$, one can find success.

Graham and Pollak (1971, 1972) proved that any connected graph has a $\{0, 1, *\}$ addressing, as in Problem 11.5.1. Given a connected graph $X = (V, E)$, **the distance multigraph** $\mathcal{D}(X)$ of $X$ is the multigraph, whose vertex set is $V$, where two vertices $x \ne y \in V$ are joined by $d(x, y)$ edges, where $d(x, y)$ denotes the distance between

**Fig. 11.7** Addressings of
some small graphs

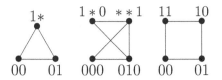

$x$ and $y$ in $X$. If $X$ is the cycle $C_4$, the distance multigraph is $C_4$ with two diagonal
double edges. If $X$ is the complete graph $K_n$, its distance multigraph is $X$ itself. A
**biclique** is a complete bipartite subgraph of $X$ (Fig. 11.7).

**Proposition 11.5.2** *Let $X = (V, E)$ be a simple and connected graph. For any
decomposition of the edge multiset of $\mathcal{D}(X)$ into $k$ bicliques, there exists an addressing of length $k$ of $X$.*

**Proof** Let $H_1, \ldots, H_k$ be a collection of bicliques whose edges decompose the edge
multiset of $\mathcal{D}(X)$. For each $1 \leq j \leq k$, let $U_j$ and $V_j$ be the partite sets of $H_j$. Define
the following function $\ell : V \rightarrow \{0, 1, *\}^k$ as follows:

$$
\ell(x)_j = \begin{cases} 0 & \text{if } x \in U_j, \\ 1 & \text{if } x \in V_j, \\ * & \text{otherwise.} \end{cases}
$$

For $x \neq y \in V$, the number of positions $j$ where $\{\ell(x)_j, \ell(y)_j\} = \{0, 1\}$ equals the
number of $H_j$, where $x \in U_j, y \in V_j$ or $x \in V_j, y \in U_j$. Since $H_1, \ldots, H_k$ decompose $\mathcal{D}(X)$, this number equals $d(x, y)$. This shows that $\ell$ is a valid addressing and
finishes the proof. ∎

Given that any graph has a valid addressing, it is interesting to determine the
smallest length $k$ of such an addressing which we will denote by $N(X)$. By the
previous proposition, $N(X)$ equals the smallest number of bicliques whose edge sets
decompose the edge multiset of $\mathcal{D}(X)$. When $X = K_n$, the complete graph with $n$
vertices, $\mathcal{D}(K_n) = K_n$ and determining $N(K_n)$ is equivalent to finding the minimum
number of bicliques whose edges partition the edge set of $K_n$.

**Theorem 11.5.3** (Graham-Pollak 1971) *If $n \geq 2$ is a natural number, then $N(K_n) =
n - 1$.*

**Proof** To prove the upper bound, assume that the vertex set of $K_n$ is $\{1, \ldots, n\}$. For
$1 \leq j \leq n - 1$, consider the star $S_j$ whose partite sets are $\{j\}$ and $\{j + 1, \ldots, n\}$,
respectively. It is not hard to see that the edge sets of $S_1, \ldots, S_{n-1}$ partition the edge
set of $K_n$. This proves that $N(K_n) \leq n - 1$.

The proof of the lower bound involves linear algebra and is due to the Norwegian mathematician Helge Tverberg (1935–2020) in Tverberg (1982). Assume
that $N(X) \leq n - 2$ and consider a decomposition of the edge set of $K_n$ into $k$
bicliques $H_1 = (U_1, V_1), \ldots, H_k = (U_k, V_k)$. Construct the following linear system

of $k + 1 \leq n - 1$ equations and $n$ variables $y_1, \ldots, y_n$ (one $y_j$ for each vertex $j$ of $K_n$):

$$y_1 + \cdots + y_n = 0$$

$$\sum_{a \in U_j} y_a = 0, \forall 1 \leq j \leq k.$$

Because we have more equations than variables, this system must have a non-zero solution $(x_1, \ldots, x_n)$. Since the edges of $H_1, \ldots, H_k$ partition the edge set of $K_n$, we deduce that

$$\sum_{1 \leq r < s \leq n} x_r x_s = \sum_{rs \in E(K_n)} x_r x_s$$

$$= \sum_{j=1}^{k} \left( \sum_{a \in U_j} x_a \right) \left( \sum_{b \in V_j} x_b \right)$$

$$= 0.$$

Combining this with $x_1 + \cdots + x_n = 0$, we get that

$$\sum_{r=1}^{n} x_r^2 = \left( \sum_{r=1}^{n} x_r \right)^2 - 2 \sum_{1 \leq r < s \leq n} x_r x_s = 0.$$

However, not all the $x_1, \ldots, x_n$ are zero means that the sum of their squares should be positive. This contradiction implies that $k \geq n - 1$ and finishes our proof. ∎

At this time, we do not know an efficient algorithm for computing $N(X)$ for a given graph $X$.

## 11.6 Exercises

**Exercise 11.6.1** Draw the line graph of $K_{3,3}$ and determine its chromatic number.

**Exercise 11.6.2** Show that the Petersen graph (Fig. 6.9) has no Hamiltonian cycles.

**Exercise 11.6.3** Show that the edge chromatic number of the Petersen graph is 4.

**Exercise 11.6.4** If $X$ is a 3-regular graph with a Hamiltonian cycle, show that $\chi'(X) = 3$.

**Exercise 11.6.5** Show that $K_{r,s}$ has no Hamiltonian cycle unless $r = s$.

**Exercise 11.6.6** Show that $K_{s,s}$ contains $\frac{s!(s-1)!}{2}$ Hamiltonian cycles.

**Exercise 11.6.7**  Show that every path of five vertices in the dodecahedron lies in a Hamiltonian cycle.

**Exercise 11.6.8**  If $X$ has a Hamiltonian path, then for each $S \subset V(X)$, the number of components of $X - S$ is at most $|S| + 1$.

**Exercise 11.6.9**  Let $n \geq 2$ be a natural number. Show that every set of $n$ integers contains a non-empty subset whose sum is divisible by $n$. Show that there are sets of $n - 1$ integers with no such subset.

**Exercise 11.6.10**  Let $m$ and $n$ be two natural numbers. Show that every sequence of $mn + 1$ real numbers contains either an increasing subsequence of length at least $m + 1$ or a decreasing subsequence of length at least $n + 1$. Show that there are sequences of $mn$ real numbers for which the above conclusion fails.

**Exercise 11.6.11**  Let $n$ and $k$ be a natural number and $a_1, \ldots, a_n$ be non-negative integers whose sum is $k$. If $k \leq 2n + 1$, show that for any $m \in [k]$, there exists $I \subset [n]$ such that

$$\sum_{i \in I} a_i = m$$

For $k = 2n + 2$, describe a set of $n$ non-negative integers for which the statement above fails.

**Exercise 11.6.12**  Let $n$ be a natural number. If $S$ is a subset $\{1, \ldots, 2n\}$ and $|S| = n + 1$, show that there are two elements in $S$ which are coprime. Describe a subset of $\{1, \ldots, 2n\}$ of size $n$ where the previous conclusion does not hold.

**Exercise 11.6.13**  Show that for any red/blue colouring of the edges of $K_6$, there exists a monochromatic cycle on four vertices. Show that this is not true for $K_5$.

**Exercise 11.6.14**  Let $n$ be a natural number. Show that the maximum number of edges in a non-Hamiltonian graph on $n$ vertices is $\binom{n-1}{2} + 1$.

**Exercise 11.6.15**  Among five points in plane such that no three of them are collinear, show there are four that determine a convex quadrilateral.

**Exercise 11.6.16**  Among nine points in plane such that no three of them collinear, show there are five that determine a convex pentagon.

**Exercise 11.6.17**  Let $n \geq 3$ be an integer number. Show that every set of $\binom{2n-4}{n-2} + 1$ points in the plane with no three collinear contains a $n$-subset, forming a convex polygon.

**Exercise 11.6.18**  Let $X$ be the 3-regular graph obtained from a cycle with 8 vertices by joining the four pairs of vertices at distance four. Prove that $\omega(X) = 2$ and $\alpha(X) = 3$.

**Exercise 11.6.19**  Show that $R(3, 4) = 9$.

**Exercise 11.6.20** A poset $P$ is a set $P$ and a binary relation $\leq$ that is reflexive, transitive, and antisymmetric. A **chain** is a sequence $a_1 < a_2 < \cdots < a_k$. An **anti-chain** is a subset of pairwise incomparable elements. If $P$ is a finite poset, show that the minimum number of chains that cover $P$ equals the maximum size of an anti-chain.

# References

R. Graham, H.O. Pollak, On the addressing problem for loop switching. Bell Syst. Tech. J. **50**, 2495–2519 (1971)

R. Graham, H.O. Pollak, On embedding graphs in squashed cubes, in *Graph Theory and Applications*. Lecture Notes in Mathematics, vol. 303 (Springer, Berlin, 1972), pp. 99–110

S. Radziszowski, Small Ramsey numbers. Electron. J. Comb. **1** (1994), Dynamic survey 1, 30 pp

H. Tverberg, On the decomposition of $K_n$ into complete bipartite graphs. J. Graph Theory **6**, 493–494 (1982)

# Chapter 12
# Expanders and Ramanujan Graphs

## 12.1 Eigenvalues and Expanders

We summarize below some basic properties of the eigenvalues of regular graphs.

**Proposition 12.1.1** *Let* $X = (V, E)$ *be a* $k$-*regular graph with* $n$ *vertices. Denote by* $\lambda_1 \geq \cdots \geq \lambda_n$ *the eigenvalues of the adjacency matrix of* $X$.

*(1) For any* $j$, $|\lambda_j| \leq k$ *and* $\lambda_1 = k$.
*(2) The graph* $X$ *is connected if and only the spectral gap* $k - \lambda_2$ *is positive.*
*(3) Assume that* $X$ *is connected. The graph* $X$ *is bipartite if and only if* $\lambda_n = -k$.

***Proof*** To see that $k$ is an eigenvalue of $A$, note that $A\vec{1} = k\vec{1}$, where $\vec{1}$ is the $n$-dimensional all one vector. Let $\lambda_j$ be some eigenvalue of $A$. Consider an eigenvector $u \in \mathbb{R}^n$ corresponding to it and denote by $x$ a vertex such that $|u_x| = \max_{y \in V} |u_y|$. Since $u$ is a non-zero vector, $|u_x| > 0$. The eigenvalue-eigenvector equation $Au = \lambda_j u$ at coordinate $x$ gives that

$$\lambda_j u_x = \sum_{y:y \sim x} u_y.$$

Taking the absolute value of both sides, using the triangle inequality and the definition of $u_x$, we obtain that

$$|\lambda_j||u_x| = |\sum_{y:y \sim x} u_y| \leq \sum_{y:y \sim x} |u_y| \leq k|u_x|,$$

which finishes the proof of part (1). For part (2), the proof is similar to the one of part (3) of Theorem 4.5.2 is omitted here. For part (3), if $X$ is bipartite, then the result is proved in Theorem 4.3.1. For the converse, assume that $v$ is a non-zero vector such that $Av = (-k)v$. Multiplying to the left by $v^T$, we get that $v^T(kI_n + A)v = v^T Av + kv^T v = 0$. This further implies that

© Hindustan Book Agency 2022
S. M. Cioabă and M. R. Murty, *A First Course in Graph Theory and Combinatorics*,
Texts and Readings in Mathematics 55, https://doi.org/10.1007/978-981-19-0957-3_12

$$0 = v^T(kI_n + A)v = \sum_{xy \in E}(v_x + v_y)^2.$$

Hence, $v_x + v_y = 0$ for any edge $xy$ of $X$. If there were a vertex $x$ with $v_x = 0$, then any vertex $z$ that can be reached by a path from $x$ would have the same property and $v_z = 0$. Since $X$ is connected, this would mean that $v$ is the zero vector, a contradiction. Hence, $v$ has no zero entries. Since $v$ is a non-zero vector, its coordinates can be split into two subsets, $V_+ = \{x \in V : v_x > 0\}$ and $V_- = \{y \in V : v_y < 0\}$. From the previous argument, there are no edges inside $V_+$ and there are no edges inside $V_-$ and hence, $X$ must be bipartite.                                                                ∎

Informally speaking, the previous result gives an intuition that for a regular graph $X$, its second-largest eigenvalue is related to its connectivity while the smallest eigenvalue is some measure of how bipartite $X$ is. It turns out that this intuition can be made formal, as follows.

For a graph $X = (V, E)$ with $n$ vertices and a subset of vertices $S$, denote by $e(S, S^c)$ the number of edges with one endpoint in $S$ and the other one in $S^c$. **The expansion constant** or **the isoperimetric constant** $h(X)$ of $X$ is defined as

$$h(X) = \min\left\{\frac{e(S, S^c)}{|S|} : S \subset V, |S| \leq n/2\right\}.$$

Informally, speaking $h(X)$ measures how connected $X$ is and if it has any *bottlenecks*. For example, for the cycle $C_n$ on $n$ vertices, it is not hard to show that $h(C_n) = 4/n$, if $n$ is even, and $h(C_n) = 4/(n-1)$ if $n$ is odd. In either case, $h(C_n)$ approaches 0 as $n$ grows.

Given $k \geq 3$, a family of connected $k$-regular graphs $(X_\ell)_{\ell \geq 1}$ is called **a family of expanders** if the following two conditions are satisfied

(1)  The number of vertices in $X_\ell$ goes to infinity as $\ell$ grows

$$\lim_{\ell \to \infty} |V(X_\ell)| = \infty. \tag{12.1.1}$$

(2)  There exists a positive constant $c > 0$ such that

$$h(X_\ell) > c, \; \forall \ell \geq 1. \tag{12.1.2}$$

Informally speaking, expanders are graphs that are both sparse (that is the meaning of condition (1)) and connected (as stated in part (2)). Expanders play an important role in computer science and the theory of communication networks. These graphs arise in questions about designing networks that connect many users while using only a few switches. Constructing families of constant-degree expanders is an important problem in mathematics and computer science. Unfortunately, computing or even approximating the expansion/isoperimetric constant of a graph is a difficult algorithmic task and there are no known efficient algorithms for it. However, the following result relating the expansion constant to the second eigenvalue of the graph helps greatly.

**Theorem 12.1.2** *If $X = (V, E)$ is a connected $k$-regular graph with adjacency matrix eigenvalues $k = \lambda_1 > \lambda_2 \geq \cdots \geq \lambda_n$, then*

$$\frac{k - \lambda_2}{2} \leq h(X) \leq \sqrt{k^2 - \lambda_2^2}. \tag{12.1.3}$$

**Proof** We prove the inequality $\frac{k-\lambda_2}{2} \leq h(X)$. Let $T$ be a subset of vertices such that $e(T, T^c)/|T| = h(X)$ and $|T| \leq n/2$. Denote $t = |T|$ and define a vector $v \in \mathbb{R}^V$, where

$$v_x = \begin{cases} n - t, & \text{if } x \in T \\ -t, & \text{if } x \in T^c. \end{cases}$$

If $\vec{1}$ denotes the $n$-dimensional all one vector, it follows that

$$\vec{1}^T v = \sum_{x \in T} v_x + \sum_{y \in T^c} v_y = t(n - t) + (n - t)(-t) = 0.$$

By an argument similar to the one in Theorem 6.4.5, it is not too hard to prove that $\lambda_2 \geq \frac{v^T A v}{v^T v}$ which implies that

$$(k - \lambda_2) v^T v \leq v^T (k I_n - A) v = \sum_{xy \in E} (v_x - v_y)^2$$

$$= e(T, T^c)(n - t + t)^2 = e(T, T^c) n^2.$$

On the other hand, $v^T v = t(n - t)^2 + (n - t)(-t)^2 = nt(n - t)$. Hence, since $t \leq n/2$, we get that

$$e(T, T^c) \geq \frac{(k - \lambda_2) t (n - t) n}{n^2} \geq \frac{(k - \lambda_2) t}{2},$$

which implies that

$$h(X) = e(T, T^c)/t \geq \frac{k - \lambda_2}{2}.$$

The proof of the other inequality is a bit more involved, and we will skip it here. The interested reader can find it in Mohar (1989, Corollary 4.5). The proof of the slightly weaker upper bound $\sqrt{2k(k - \lambda_2)}$ for $h(X)$ can be found in Hoory et al. (2006, Theorem 2.4). ∎

This theorem implies that to construct an infinite family of $k$-regular expanders $(X_\ell)_{\ell \geq 1}$, the part (2) from the definition can be replaced by the condition that there exists some constant $c' > 0$ such that $k - \lambda_2(X_\ell) > c'$. This is equivalent to $\lambda_2(X_\ell) < k - c'$ which means that we should keep $\lambda_2(X_\ell)$ as far away from $k$ as possible. The eigenvalues of a graph can be computed efficiently, and the problem of constructing expanders becomes the problem of constructing $k$-regular graphs with *small* $\lambda_2$.

In communication theory, one requires the networks to have small diameter for efficient operation. We mention now a property of expanders that is also useful, namely, the fact that they have *small* diameter. Note that the diameter $\text{diam}(X)$ of a connected $k$-regular graph $X$ grows at least as the logarithm of the number of vertices of $X$. When $k \geq 3$, one can prove that $\text{diam}(X) \geq \frac{\log(n-1)}{\log(k-1)} + \log_{k-1}((k-2)/2)$ (see Exercise 5.6.19). Here is a weaker bound with a shorter proof. If $d = \text{diam}(X)$, then the number of paths of length $d$ starting at a given vertex $x$, is at least $k^d$ and as each such path has $d + 1$ vertices, we deduce that $n \leq (d + 1)k^d$. This implies our assertion about the growth of $\text{diam}(X)$.

One can also prove that expanders asymptotically reach this bound and have logarithmic diameter. We leave the details to the reader and present now an estimate for the diameter due to Chung (1989). We need one notation. Let $X$ be a connected $k$-regular graph with $n$ vertices. If $k = \lambda_1 > \lambda_2 \geq \cdots \geq \lambda_n$ are the eigenvalues of $X$, define

$$\lambda = \begin{cases} \max_{2 \leq j \leq n} |\lambda_j|, & \text{if } X \text{ is not bipartite,} \\ \max_{2 \leq j \leq n-1} |\lambda_j|, & \text{if } X \text{ is bipartite.} \end{cases}$$

**Theorem 12.1.3** *Let $X = (V, E)$ be a connected $k$-regular graph with $n$ vertices.*

*(1) If $X$ is not bipartite, then*

$$\text{diam}(X) \leq \left\lceil \frac{\log(n - 1)}{\log(k/\lambda)} \right\rceil.$$

*(2) If $X$ is bipartite, then*

$$\text{diam}(X) \leq \left\lceil \frac{\log((n - 1)/2)}{\log(k/\lambda)} \right\rceil.$$

***Proof*** We give the proof of (1) and leave the one of part (2) as an exercise for the motivated reader. Let $A$ be the adjacency matrix of $X$. For $x, y \in V$, the $(x, y)$th entry of $A^r$ is the number of $(x, y)$-walks of length $r$. We find a natural number $r$ such that all the entries of $A^r$ are positive. By the previous remark, this implies that $\text{diam}(X) \leq r$.

Let $n = |V|$ and $u_1, u_2, \ldots, u_n$ be an orthonormal basis of $\mathbb{R}^n$ formed by eigenvectors of $A$ with corresponding eigenvalues, $k = \lambda_1, \lambda_2, \ldots, \lambda_n$, respectively. We may take $u_1 = \vec{1}/\sqrt{n}$ where $\vec{1}$ is the all one $n$-dimensional vector. The spectral decomposition theorem tells that

$$A^r = \sum_{i=1}^{n} \lambda_i^r u_i u_i^T,$$

for any $r \geq 1$. Hence, for any $x, y \in V$, using the Cauchy-Schwarz inequality, we get that

$$
\begin{aligned}
A^r(x, y) &= \sum_{i=1}^{n} \lambda_i^r (u_i u_i^T)_{x,y} \\
&\geq \frac{k^r}{n} - \left| \sum_{i=2}^{n} \lambda_i^r (u_i)_x (u_i)_y \right| \\
&\geq \frac{k^r}{n} - \lambda^r \left( \sum_{i=2}^{n} (u_i)_x^2 \right)^{1/2} \left( \sum_{i=1}^{n} (u_i)_y^2 \right)^{1/2} \\
&= \frac{k^r}{n} - \lambda^r (1 - (u_0)_x^2)^{1/2} (1 - (u_0)_y^2)^{1/2} \\
&= \frac{k^r}{n} - \lambda^r (1 - 1/n).
\end{aligned}
$$

This inequality implies that $A^r(x, y) > 0$ for any $x, y \in V$ whenever $\frac{k^r}{\lambda^r} > n - 1$. If we take $r = \left\lceil \frac{\log(n-1)}{\log(k/\lambda)} \right\rceil$, this finishes our proof. ∎

## 12.2   The Alon-Boppana Theorem

A natural question is how small can $\lambda_2$ be for a connected $k$-regular graph or for a sequence of connected $k$-regular graphs whose number of vertices is growing larger. The answer is given by the Alon-Boppana theorem, proved in the mid 1980s by Noga Alon and Ravi Boppana.

The first appearance in print of this result that we are aware of, is in 1986 in Alon's paper (Alon 1986) where it is stated that

*R. Boppana and the present author showed that for every d-regular graph G on n vertices $\lambda(G) \leq d - 2\sqrt{d-1} + O(\log_d n)^{-1}$.*

Note that the $\lambda$ in Alon (1986) is the smallest positive eigenvalue of the Laplacian $D - A = dI - A$ of a $d$-regular graph $G$ and it equals $d - \lambda_2(G)$. Therefore, the statement above is equivalent to

$$
\lambda_2(G) \geq 2\sqrt{d-1} - O(\log_d^{-1} n). \tag{12.2.1}
$$

The first proof of this result appeared in print in 1991 in the paper (Nilli 1991) by A. Nilli (which is pseudonym of Noga Alon). Note that there is a proof of Alon-Boppana theorem in an earlier paper of Lubotzky et al. (1988) with a weaker error term

$$
\lambda_2(G) \geq 2\sqrt{d-1} \left( 1 - \frac{c \log D}{D} \right) \tag{12.2.2}
$$

where $c > 0$ is some constant and $D$ is the diameter of the graph. The proof we give below is from a paper by Nilli (1991).

**Theorem 12.2.1** (Alon-Boppana) *Let $X = (V, E)$ be a connected $k$-regular graph in which there are two edges at distance at least $2t + 2 \geq 4$. Then*

$$\lambda_2 \geq 2\sqrt{k-1} - \frac{2\sqrt{k-1}-1}{t+1}.$$

**Proof** Let $u_1 u_2$ and $v_1 v_2$ be two edges at distance at least $2t + 2$, meaning that

$$\min_{1 \leq i,j \leq 2} d(u_i, v_j) \geq 2t + 2. \qquad (12.2.3)$$

For $0 \leq i \leq t$, define

$$U_i = \{x : \min(d(x, u_1), d(x, u_2)) = i\}$$
$$V_i = \{y : \min(d(y, v_1), d(y, v_2)) = i\}.$$

Assumption (12.2.3) implies that $U_0 \cup \cdots \cup U_t$ is disjoint from $V_0 \cup \cdots \cup V_t$ and there are no edges between them. Also, for $1 \leq i \leq t$,

$$|U_i| \leq (k-1)|U_{i-1}|$$
$$|V_i| \leq (k-1)|V_{i-1}|.$$

This is the place where the fact that we use the distance partition from an edge (instead from a vertex) comes into play, namely, that $|U_1| \leq (k-1)|U_0|$ and $|V_1| \leq (k-1)|V_0|$. If $U_0$ were a single vertex, then we would have $|U_1| \leq k|U_0|$ instead.

Because $\lambda_2 = \max_{f \perp \vec{1}, f \neq \vec{0}} \frac{f^T A f}{f^T f}$, we have to come up with an appropriate vector $f$ to get a good lower bound for $\lambda_2$. Let $a > 0 > b$ be two constants to be chosen later and take $f : V \to \mathbb{R}$ as follows:

$$f(w) = \begin{cases} a(k-1)^{-i/2}, & \text{if } w \in U_i, 0 \leq i \leq t \\ b(k-1)^{-i/2}, & \text{if } w \in V_i, 0 \leq i \leq t \\ 0, & \text{otherwise.} \end{cases} \qquad (12.2.4)$$

The condition $\vec{1} \perp f$ is equivalent to

$$a \sum_{i=0}^{k} |U_i|(d-1)^{-i/2} + b \sum_{i=0}^{k} |V_i|(d-1)^{-i/2} = 0$$

so we may take $a = 1$ and

$$b = -\frac{\sum_{i=0}^{t} |V_i|(k-1)^{-i/2}}{\sum_{i=0}^{t} |U_i|(k-1)^{-i/2}}.$$

Note that $f^T f = A_1 + B_1$, where

$$A_1 = a^2 \sum_{i=0}^{t} \frac{|U_i|}{(k-1)^i}$$

$$B_1 = b^2 \sum_{i=0}^{t} \frac{|V_i|}{(k-1)^i}.$$

We also have that

$$f^T(I_n - A)f = \sum_{xy \in E} (f(x) - f(y))^2 = A_2 + B_2,$$

where

$$A_2 = a^2 \sum_{i=0}^{t-1} e(U_i, U_{i+1}) \left((k-1)^{-i/2} - (k-1)^{-(i+1)/2}\right)^2$$
$$+ a^2 e(U_t, U_{t+1})(k-1)^{-1/t},$$

and

$$B_2 = b^2 \sum_{i=0}^{t-1} e(V_i, V_{i+1}) \left((k-1)^{-i/2} - (k-1)^{-(i+1)/2}\right)^2$$
$$+ b^2 e(V_t, V_{t+1})(k-1)^{-1/t},$$

where $U_{k+1} = V_{k+1} = V \setminus (\cup_{i=0}^{t}(U_i \cup V_i))$.

Putting these equations together, we get that

$$k - \lambda_2 \le \frac{f^T(kI - A)f}{f^T f} = \frac{A_2 + B_2}{A_1 + B_1} \le \max\left(A_2/A_1, B_2/B_1\right). \qquad (12.2.5)$$

Because $e(U_i, U_{i+1}) \le (k-1)|U_i|$ for $0 \le i \le t - 1$, we deduce that

$$A_2 \le (k - 2\sqrt{k-1})A_1 + a^2 \frac{|U_t|(2\sqrt{k-1} - 1)}{(k-1)^t}.$$

Now because $(k-1)|U_i| \ge |U_{i+1}|$ for $0 \le i \le t - 1$, we deduce that

$$\frac{|U_0|}{(k-1)^0} \ge \frac{|U_1|}{(k-1)^1} \ge \cdots \ge \frac{|U_t|}{(k-1)^t}$$

and therefore

$$A_1 = a^2 \sum_{i=0}^{t} \frac{|U_i|}{(k-1)^i} \ge (t+1)a^2 \frac{|U_t|}{(k-1)^t}.$$

Thus, $a^2 \frac{|U_t|}{(k-1)^t} \le \frac{A_1}{t+1}$. Plugging this inequality in the inequality above involving $A_2$ and $A_1$, we obtain that

$$A_2 \le (k - 2\sqrt{k-1})A_1 + \frac{(2\sqrt{k-1}-1)A_1}{t+1}$$

which means that

$$A_2/A_1 \le k - 2\sqrt{k-1} + \frac{k - 2\sqrt{k-1}}{t+1}.$$

By a similar argument, one can show that

$$B_2/B_1 \le k - 2\sqrt{k-1} + \frac{k - 2\sqrt{k-1}}{k+1}.$$

Using (12.2.5), we obtain that

$$k - \lambda_2 \le k - 2\sqrt{k-1} + \frac{2\sqrt{k-1}-1}{t+1}$$

which implies the desired inequality

$$\lambda_2 \ge 2\sqrt{k-1} - \frac{2\sqrt{k-1}-1}{t+1}$$

and finishes our proof. ∎

For a connected $k$-regular graph $X$ of diameter $\mathrm{diam}(X)$, we can find two pairs of vertices as in the hypothesis of the previous theorem for $t$ about $\lfloor \mathrm{diam}(X)/2 \rfloor - 1$. Since the diameter of a regular graph grows at least as the logarithm of the order of the graph (see the end of the previous section), the Alon-Boppana Theorem has the following important consequence.

**Theorem 12.2.2** (Asymptotic Alon-Boppana) *Let $k \ge 3$ be a natural number. If $(X_\ell)_{\ell \ge 1}$ is a sequence of connected $k$-regular graphs such that*

$$\lim_{\ell \to \infty} |V(X_\ell)| = \infty,$$

*then*

$$\liminf_{\ell \to \infty} \lambda_2(X_\ell) \ge 2\sqrt{k-1}.$$

There is a similar result to the Alon-Boppana theorem due to Serre (1997). The meaning of the theorem below is that large $k$-regular graphs tend to have a positive proportion of eigenvalues trying to be greater than $2\sqrt{k-1}$ (and not just $\lambda_2$). The proof given below is from Cioabă (2006).

**Theorem 12.2.3** *For any $k \geq 3$, $\epsilon > 0$, there exists $c = c(\epsilon, k) > 0$ such that any $k$-regular graph $X$ with $n$ vertices has at least $c \cdot n$ eigenvalues that are at least $2\sqrt{k-1} - \epsilon$.*

**Proof** Let $k \geq 3$ be a natural number and let $\epsilon > 0$ be a positive number. Consider a $k$-regular graph $X$ with adjacency matrix $A$ and eigenvalues $k = \lambda_1 \geq \lambda_2 \geq \cdots \geq \lambda_n$. We use the following notation, $m := |\{\lambda_j : \lambda_j \geq 2\sqrt{k-1} - \epsilon\}|$ which obviously means that $n - m = |\{\lambda_j : \lambda_j < 2\sqrt{k-1} - \epsilon\}|$. From these inequalities, we deduce that

$$
\begin{aligned}
\operatorname{tr}\left[(kI + A)^{2s}\right] &= \sum_{i=1}^{n}(k + \lambda_i)^{2s} \\
&< (n - m)(k + 2\sqrt{k-1} - \epsilon)^{2s} + m(2k)^{2s} \\
&= m[(2k^{2s} - (k + 2\sqrt{k-1} - \epsilon)^{2s}] \\
&\quad + n(k + 2\sqrt{k-1} - \epsilon)^{2s}.
\end{aligned}
$$

The number of closed walks of length $2r$ in $G$ starting at some vertex $v$ is at least the number of closed walks of length $2r$ in the infinite $k$-regular tree starting at its root. It can be shown (see Cioabă 2006; McKay 1981) that this number is at least $\frac{1}{r+1}\binom{2r}{r}k(k-1)^{r-1}$. Because the trace of $A^{2r}$ equals the number of closed walks of length $2r$ in $G$, we deduce that

$$
\operatorname{tr}(A^{2r}) > \frac{n}{r+1}\binom{2r}{r}k(k-1)^{r-1} > \frac{n(2\sqrt{k-1})^{2r}}{(r+1)^2},
$$

where in the second inequality above we used that $\binom{2r}{r} > \frac{4^r}{r+1}$ for any $r \geq 1$ (this can be proved by induction on $r$ for example). Using the inequality above and binomial expansion, we get that

$$
\begin{aligned}
\operatorname{tr}\left[(kI + A)^{2s}\right] &= \sum_{i=0}^{2s}\binom{2s}{i}k^i \operatorname{tr}(A^{2s-i}) \\
&\geq \sum_{j=0}^{s}\binom{2s}{2j}k^{2j}\operatorname{tr}(A^{2s-2j}) \\
&> \frac{n}{(s+1)^2}\sum_{j=0}^{s}\binom{2s}{2j}k^{2j}(2\sqrt{k-1})^{2s-2j} \\
&= \frac{n}{2(s+1)^2}\left[(k + 2\sqrt{k-1})^{2s} + (k - 2\sqrt{k-1})^{2s}\right]
\end{aligned}
$$

$$> \frac{n}{2(s+1)^2}(k+2\sqrt{k-1})^{2s}.$$

Combining the above upper and the lower bound for $\mathrm{tr}\,[(kI+A)^{2s}]$, we obtain the following:

$$\frac{m}{n} > \frac{\frac{(k+2\sqrt{k-1})^{2s}}{(s+1)^2} - (k+2\sqrt{k-1}-\epsilon)^{2s}}{(2k)^{2s} - (k+2\sqrt{k-1}-\epsilon)^{2s}}.$$

Because

$$\lim_{s\to\infty}\left(\frac{(k+2\sqrt{k-1})^{2s}}{2(s+1)^2}\right)^{\frac{1}{2s}} = k+2\sqrt{k-1}$$

$$> k+2\sqrt{k-1}-\epsilon$$

$$= \lim_{s\to\infty}(2(k+2\sqrt{k-1}-\epsilon)^{2s})^{\frac{1}{2s}},$$

we deduce that there exists $s_0 = s_0(\epsilon, d)$ such that for any $s \geq s_0$,

$$\frac{(k+2\sqrt{k-1})^{2s}}{2(s+1)^2} - (k+2\sqrt{k-1}-\epsilon)^{2s} > (k+2\sqrt{k-1}-\epsilon)^{2s}. \qquad (12.2.6)$$

If we take $c(\epsilon, k) = \frac{(k+2\sqrt{k-1}-\epsilon)^{2s_0}}{(2k)^{2s_0}-(k+2\sqrt{k-1}-\epsilon)^{2s_0}}$, then $c(\epsilon, k) > 0$, and $m > c(\epsilon, k)n$. ∎

There is a similar result for the smallest eigenvalues of regular graphs, whose proof we skip here and invite the reader to complete.

**Theorem 12.2.4** (Cioabă 2006) *For any $\epsilon > 0$ and $k \geq 3$, there exist a positive constant $c = c(\epsilon, k) > 0$ and a natural number $g = g(\epsilon, k)$ such that any $k$-regular graph on $n$ vertices without any odd cycles of length less than $g$ must have at least $c \cdot n$ eigenvalues that are less than $-2\sqrt{k-1}+\epsilon$.*

These results mean that $2\sqrt{k-1}$ is the best bound one can hope to obtain for the absolute value of the non-trivial eigenvalues of a $k$-regular graph. This motivates the following definition that appeared first in the paper (Lubotzky et al. 1988) by Alex Lubotzky, Ralph Phillips and Peter Sarnak. Let $k \geq 3$ be a natural number. A connected $k$-regular graph $X$ is called **Ramanujan** if

$$|\lambda_j| \leq 2\sqrt{k-1},$$

for any eigenvalue $\lambda_j$ of $X$ except $k$ and possibly $-k$ (if $X$ is bipartite). Their paper (Lubotzky et al. 1988) gave the first explicit infinite family of $k$-regular Ramanujan graphs when $k-1$ is prime. This result was also obtained independently by Margulis (1988).

As these constructions used the theory of modular forms and the Ramanujan estimate for eigenvalues of Hecke operators occurring in this theory, Lubotzky, Phillips and Sarnak named these graphs after the famous Indian mathematician Srinivasa Ramanujan (1887–1920). We describe their construction in Sect. 12.5.

Given $k \geq 3$, it may not be too hard to find some small examples of $k$-regular Ramanujan graphs. We leave it for the reader to verify the validity of the following statements.

**Example 12.2.5** For any $k \geq 3$, the complete graph $K_{k+1}$ is Ramanujan.

**Example 12.2.6** For any $k \geq 3$, the complete bipartite graph $K_{k,k}$ is Ramanujan.

**Example 12.2.7** The Petersen graph is Ramanujan.

However, the explicit construction of an infinite family of $k$-regular Ramanujan graphs is a much more difficult problem which is of great importance in mathematics and computer science. As noted above, the first such construction was obtained by Lubotzky et al. (1988) and independently, Margulis (1988). As mentioned earlier and described in Sect. 12.5, these authors used deep results from algebra, number theory and combinatorics to construct infinite families of $k$-regular Ramanujan graphs, whenever $k - 1$ is a prime (see also Davidoff et al. 2003 for a nice exposition of this construction). Their work was extended by Morgenstern (1994) who constructed infinite families of $k$-regular Ramanujan graphs when $k - 1$ is a prime power. More recently, Marcus et al. (2015) proved that for any $k \geq 3$, there exists an infinite family of $k$-regular bipartite Ramanujan graphs. Their result is existential and, so far, nobody has managed to extend it to show the existence of infinite $k$-regular non-bipartite Ramanujan graphs for any $k \geq 3$. Also, constructing explicit families of such graphs is open for the values of $k$ where $k - 1$ is not a prime power. For example, no one has been able to construct an infinite family of 7-regular Ramanujan graphs. We describe the ideas behind their work in Sect. 12.5. The student can find a survey of recent developments in Ram Murty (2020).

## 12.3 Group Characters and Cayley Graphs

**A character** $\chi$ of a group $G$ is a group homomorphism from $G$ to the multiplicative group $\mathbb{C}^*$ of non-zero complex numbers. This means that

$$\chi(ab) = \chi(a)\chi(b), \forall a, b \in G. \tag{12.3.1}$$

The character $\chi_0$ that sends every element to the identity element 1 of $\mathbb{C}^*$ is called the **trivial character**.

The basic idea of character theory is that to understand the abstract group $G$, we map it into something concrete like the multiplicative group of complex numbers and see how the image looks like to deduce what $G$ looks like. We summarize the basic properties of the characters of a finite group below.

**Proposition 12.3.1** *Let $G$ be a group of order $n$ and let $\chi$ be a character of $G$.*

*(1)* $\chi(1) = 1$.
*(2) If $a \in G$, then $\chi(a^{-1}) = \chi(a)^{-1}$.*
*(3) If $a \in G$, then $\chi(a)$ is a nth root of unity, $\chi(a)^n = 1$.*

***Proof*** For part (1), taking $a = b = 1$ in Eq. (12.3.1), we get that $\chi(1) = \chi(1)^2$. Since $\chi(1) \neq 0$, this means that $\chi(1) = 1$.

For part (2), let $a \in G$. Taking $b = a^{-1}$ in (12.3.1), we obtain that $\chi(1) = \chi(a)\chi(a^{-1})$. This implies the desired result.

For part (3), recall that Lagrange's theorem (see Theorem 7.1.11) implies that $a^n = 1$. Combining with Eq. (12.3.1), we deduce that $1 = \chi(1) = \chi(a^n) = \chi(a)^n$ which is the result we wanted to prove. ∎

**Example 12.3.2** For the additive group $(\mathbb{Z}_n, +)$ of residue classes mod $n$, the characters are $\chi_0, \chi_1, \ldots, \chi_{n-1}$ given by

$$\chi_j(a) = e^{2\pi i j a/n}, a \in \mathbb{Z}_n, \tag{12.3.2}$$

for $0 \leq j \leq n - 1$. Notice that $\chi_0$ is again the trivial character.

The characters of a finite Abelian group can be determined using the following result from group theory.

**Theorem 12.3.3** *If $G$ is an Abelian group of order $n$, then there exist a natural number $s \geq 1$, primes $p_1, \ldots, p_s$ (not necessarily distinct) and natural numbers $a_1, \ldots, a_s$ such that*

$$G \cong \times_{j=1}^s (\mathbb{Z}_{p_j^{a_j}}, +).$$

**Example 12.3.4** If $p$ is a prime, there is only one Abelian group of order $p$, up to isomorphism, namely $(\mathbb{Z}_p, +)$.

**Example 12.3.5** If $p$ is a prime, then there are exactly two Abelian groups of order $p^2$, namely $(\mathbb{Z}_{p^2}, +)$ and $(\mathbb{Z}_p, +) \times (\mathbb{Z}_p, +)$.

**Example 12.3.6** There are exactly three Abelian groups of order 8:

$$(\mathbb{Z}_8, +), (\mathbb{Z}_4 \times \mathbb{Z}_2, +), (\mathbb{Z}_2 \times \mathbb{Z}_2 \times \mathbb{Z}_2, +).$$

For a group $G$, the set of characters in turn forms a group under multiplication of characters. Indeed, we define for two characters $\chi$ and $\psi$, **the product**

$$(\chi\psi)(a) := \chi(a)\psi(a), \forall a \in G,$$

is also a character of $G$. The characters form a group called **the character group of $G$**, usually denoted by $\hat{G}$. The identity element of $\hat{G}$ is the trivial character. The character inverse to $\chi$ is $\chi^{-1}$ defined by

$$\chi^{-1}(a) = \chi(a)^{-1}, \forall a \in G.$$

With these results in mind, the characters of a finite Abelian group of the form $\times_{j=1}^{s}(\mathbb{Z}_{p_j^{a_j}}, +)$ are the products of the characters of the groups $(\mathbb{Z}_{p_1^{a_1}}, +), \ldots,$ $(\mathbb{Z}_{p_s^{a_s}}, +)$.

The following is a simple and elegant procedure for constructing $k$-regular graphs using group theory. Let $G$ be a finite group and $S$ a $k$-element subset of $G$. We suppose that $S$ **symmetric**, meaning that $s \in S$ implies $s^{-1} \in S$. Construct the graph $X(G, S)$, whose vertex set is $G$, and where $(x, y)$ is an edge if and only if $xy^{-1} \in S$. The graph $X(G, S)$ is called **the Cayley graph of the group $G$ with generating set $S$** and is regular with valency $|S|$. The graph $X(G, S)$ will contain loops if and only if the identity element 1 of $G$ is contained in $S$. If the group $G$ is Abelian, the eigenvalues of the Cayley graph are easily determined as follows.

**Theorem 12.3.7** *Let $G$ be a finite Abelian group and $S$ a symmetric subset of $G$ of size $k$. The eigenvalues of the adjacency matrix of $X(G, S)$ are given by*

$$\lambda_\chi = \sum_{s \in S} \chi(s),$$

*where $\chi$ ranges over all the characters of $G$.*

**Proof** For each irreducible character $\chi$, let $v_\chi$ denote the vector $(\chi(g) : g \in G)$. Let $\delta_S(g)$ equal 1 if $g \in S$ and zero otherwise, and denote by $A$ the adjacency matrix of $X(G, S)$. If $x \in G$, then

$$(Av_\chi)_x = \sum_{g \in S} \delta_S(xg^{-1})\chi(g).$$

By replacing $xg^{-1}$ by $s$, and using the fact that $S$ is symmetric, we obtain

$$(Av_\chi)_x = \chi(x)\left(\sum_{s \in S} \chi(s)\right)$$

which shows that $v_\chi$ is an eigenvector with eigenvalue

$$\sum_{s \in S} \chi(s).$$

Note that any two distinct eigenvectors described above are orthogonal and therefore, the spectrum of $X(G, S)$ consists of the eigenvalues $\lambda_\chi$.  ∎

Notice that for the trivial character, we have $\lambda_1 = k$. Thus, to construct Ramanujan graphs, we require

$$\left|\sum_{s \in S} \chi(s)\right| \leq 2\sqrt{k-1}$$

for every character $\chi$ of $G$.

The previous calculation is reminiscent of the Dedekind determinant formula in number theory, which computes det $A$ where $A$ is the matrix whose $(i, j)$th entry is $f(ij^{-1})$ for any function $f$ defined on the finite Abelian group $G$ of order $n$. The determinant is

$$\prod_\chi \left( \sum_{g \in G} f(g)\chi(g) \right).$$

The proof is analogous to the calculation in the proof of Theorem 12.3.7. As an application, it allows us to compute the determinant of a circulant matrix. For instance, we can compute the characteristic polynomial of the complete graph. Indeed, it is not hard to see that by taking the additive cyclic group of order $n$ and setting $f(0) = -\lambda$, $f(a) = 1$ for $a \neq 0$, we obtain that the characteristic polynomial is $(-1)^n(\lambda - (n-1))(\lambda - 1)^{n-1}$ by the Dedekind determinant formula.

If $G$ is an Abelian group and $S$ is a subset of $G$, we can define another set of graphs $Y(G, S)$ called **sum graphs** as follows. The vertices consist of elements of $G$ and $(x, y)$ is an edge if $xy \in S$.

**Theorem 12.3.8** *Let $G$ be an Abelian group. For each character $\chi$ of $G$, the eigenvalues of $Y(G, S)$ are given as follows. Define*

$$e_\chi = \sum_{s \in S} \chi(s).$$

*If $e_\chi = 0$, then $v_\chi$ and $v_\chi^{-1}$ are both eigenvectors with eigenvalues zero. If $e_\chi \neq 0$, then*

$$|e_\chi|v_\chi \pm e_\chi v_{\chi^{-1}}$$

*are two eigenvectors with eigenvalues $\pm|e_\chi|$.*

Using this theorem, Winnie Li constructed Ramanujan graphs in the following way. Let $\mathbb{F}_q$ denote the finite field of $q$ elements. Let $G = \mathbb{F}_{q^2}$ and take for $S$ the elements of $G$ of norm one. This is a symmetric subset of $G$ and the Cayley graph $X(G, S)$ turns out to be Ramanujan. The latter is a consequence of a theorem of Deligne estimating Kloosterman sums.

There is a generalization of these results to the non-Abelian context. This is essentially contained in a paper by Diaconis and Shahshahani (1981). Using their results, one can easily generalize the Dedekind determinant formula as follows (and which does not seem to be widely known). Let $G$ be a finite group and $f$ a class function on $G$. Then the determinant of the matrix $A$ whose rows (and columns) are indexed by the elements of $G$ and whose $(i, j)$th entry is $f(ij^{-1})$ is given by

$$\prod_\chi \left( \frac{1}{\chi(1)} \sum_{g \in G} f(g)\chi(g) \right)^{\chi(1)}$$

with the product over the distinct irreducible characters of $G$.

**Theorem 12.3.9** *Let $G$ be a finite group and $S$ a subset which is stable under conjugation. The eigenvalues of the adjacency matrix of the Cayley graph $X(G, S)$ are given by*

$$\lambda_\chi = \frac{1}{\chi(1)} \sum_{s \in S} \chi(s)$$

*as $\chi$ ranges over all irreducible characters of $G$. Moreover, the multiplicity of $\lambda_\chi$ is $\chi(1)^2$.*

We remark that the $\lambda_\chi$ in the above theorem need not be all distinct. For example, if there is a non-trivial character $\chi$ which is trivial on $S$, then the multiplicity of the eigenvalue $|S|$ is at least $1 + \chi(1)^2$.

*Proof* We essentially modify the proof of Diaconis and Shahshahani to suit our context. We consider the group algebra $\mathbb{C}[G]$ with basis vectors $e_g$ with $g \in G$ and multiplication defined as usual by $e_g e_h = e_{gh}$. We define the linear operator $Q$ by

$$Q = \sum_{s \in S} e_s = \sum_{g \in G} \delta_S(g) e_g$$

which acts on $\mathbb{C}[G]$ by left multiplication. The matrix representation of $Q$ with respect to the basis vectors $e_g$ with $g \in G$ is precisely the adjacency matrix of $X(G, S)$ as is easily checked. If $r$ denotes the left regular representation of $G$ on $\mathbb{C}[G]$, we find that the action of

$$r(A) = \sum_{s \in S} r(s)$$

on $\mathbb{C}[G]$ is identical to $Q$. Moreover, $\mathbb{C}[G]$ decomposes as

$$\mathbb{C}[G] = \oplus_\rho V_\rho,$$

where the direct sum is over non-equivalent irreducible representations of $G$ and the subspace $V_\rho$ is a direct sum of $\deg \rho$ copies of the subspace $W_\rho$ corresponding to the irreducible representation $\rho$. The result is now clear from basic facts of linear algebra. ∎

## 12.4 The Ihara Zeta Function of a Graph

Let $X = (V, E)$ be a graph with adjacency matrix $A$. As we have seen earlier, for any $x, y \in V$ and natural number $r$, $A^r(x, y)$ equals the numbers of $(x, y)$-walks of length $r$.

A walk $x = x_0, x_1, \ldots, x_r = y$ is called **a nonbacktracking walk** or a **nonbacktracking** $(x, y)$**-walk** if $x_{j-1} \neq x_{j+1}$ for $1 \leq j \leq r - 1$. Informally, this means that after we walk from $x_{j-1}$ to $x_j$ on the edge $x_{j-1}x_j$, we cannot use the same edge $x_jx_{j-1}$ at the next step, although this edge may be used later in the walk.

Let $A_r$ denote the matrix whose $(x, y)$th entry will be the number of nonbacktracking $(x, y)$-walk of length $r$. It turns out that for regular graphs, $A_r$ is a polynomial in $A$, the adjacency matrix of $X$.

**Proposition 12.4.1** *Let $X$ be a connected $k$-regular graph with adjacency matrix $A$. Define a sequence of polynomials $(f_r(X))_{r \geq 0}$ as follows: $f_0(X) = 1$, $f_1(X) = X$, $f_2(X) = X^2 - k$ and*

$$f_{r+1}(X) = Xf_r(X) - (k-1)f_{r-1}, \forall r \geq 2.$$

*For any $r \geq 0$,*
$$A_r = f_r(A).$$

**Proof** Note that $A_0 = I$, $A_1 = A$ and $A^2 = A_2 + kI_n$. Also, for any $r \geq 2$, we have that

$$A_1A_r = A_{r+1} + (k-1)A_{r-1}. \tag{12.4.1}$$

Since the left-hand side counts the number of walks of length $r + 1$ which are extended from nonbacktracking walks of length $r$ and the right-hand side enumerates first the nonbacktracking walks of length $r + 1$ and the nonbacktracking walks of length $r - 1$ which are extended to walks of length $r + 1$ with a backtracking. The conclusion follows using strong induction on $r$. ∎

The previous recursion allows us to deduce the following identity of formal power series.

**Proposition 12.4.2** *If $X$ is a connected $k$-regular graph with $n$ vertices, adjacency matrix $A$ and nonbacktracking walk matrices $A_r, r \geq 0$, then*

$$\left(\sum_{r=0}^{\infty} A_r t^r\right)(I_n - At + (k-1)t^2) = (1-t^2)I_n.$$

Let $X$ be a $k$-regular graph and set $q = k - 1$. Motivated by the theory of the Selberg zeta function, Ihara was led to make the following definitions and construct the graph-theoretic analogue of it as follows. A proper path whose endpoints are equal is called a *closed geodesic*. If $\gamma$ is a closed geodesic, we denote by $\gamma^r$ the closed geodesic obtained by repeating the path $\gamma$ $r$ times. A closed geodesic which is not the power of another one is called a *prime geodesic*. We define an equivalence relation on the closed geodesics $(x_0, \ldots, x_n)$ and $(y_0, \ldots, y_m)$ if and only if $m = n$ and there is a $d$ such that $y_i = x_{i+d}$ for all $i$ (and the subscripts are interpreted modulo $n$. An equivalence class of a prime geodesic is called a *prime geodesic cycle*. Ihara then defines the zeta function

$$Z_X(s) = \prod_p \left(1 - q^{-s\ell(p)}\right)^{-1},$$

where the product is over all prime geodesic cycles and $\ell(p)$ is the length of $p$. Ihara proves the following theorem.

**Theorem 12.4.3** *For* $g = (q - 1)|X|/2$,

$$Z_X(s) = (1 - u^2)^{-g} \det(I_n - Au + qu^2)^{-1}, \quad u = q^{-s}.$$

*Moreover,* $Z_X(s)$ *satisfies* the Riemann hypothesis, *that is, all the singular points lie on the line* $Re(s) = 1/2$ *if and only if* X *is a Ramanujan graph.*

**Proof** *(Sketch)* We assume that the zeta function has the shape given and show that it satisfies the Riemann hypothesis if and only if $X$ is Ramanujan. Let $\phi(z) = \det(zI_n - A)$ be the characteristic polynomial of $A$. If we set $z = (1 + qu^2)/u$, then the singular points of the $Z_X(s)$ arise from the zeros of $\phi(z)$. Since

$$u = \frac{z \pm \sqrt{z^2 - 4q}}{2q}$$

and any zero of $\phi$ is real (because $A$ is symmetric), we deduce that

$$\frac{z\bar{u}}{\bar{u}} = \frac{(1 + qu^2)\bar{u}}{u\bar{u}} = \frac{\bar{u} + q|u|^2 u}{|u|^2}$$

is also real. Thus, the numerator is real and so, we must have

$$q|u|^2 = 1,$$

which is equivalent to the assertion of the theorem. ∎

## 12.5 Ramanujan Graphs

The first explicit constructions of Ramanujan graphs were given (independently) by Margulis (1988) and Lubotzky et al. (1988), almost 20 years after Ihara's work. They constructed infinite families of Ramanujan graphs of degree $p + 1$ with $p$ prime. The proof that their graphs were Ramanujan relied on the analogue of Ramanujan's conjecture for cusp forms of weight 2 (proved by Eichler and Shimura) and this explains the name, Ramanujan graphs, to some extent. Later, in 1994, Morgenstern (1994) constructed explicitly infinite families of Ramanujan graphs of degree $q + 1$ when $q$ is any prime power. His work used the analogue of Ramanujan's conjecture for Drinfeld modules, proved by Drinfeld (1988). No explicit construction of infinite families of Ramanujan graphs for other degrees is known.

Here is a short description of the explicit construction of Ramanujan graphs of degree $p + 1$ for every prime $p$ given in Lubotzky et al. (1988). Let $p$ and $q$ be unequal primes, with both being $\equiv 1 \pmod 4$. By elementary number theory, the congruence $u^2 \equiv -1 \pmod q$ has an integer solution. Fix any such solution. Now, by classical theorems of Lagrange and Jacobi, the equation $p = a^2 + b^2 + c^2 + d^2$ has precisely $8(p + 1)$ integer solutions. Among these, there are exactly $p + 1$ solutions with $a > 0$ and $b, c, d$, even. To each such solution, we associate the matrix:

$$\begin{bmatrix} a + ub & c + ud \\ -c + ud & a - ub \end{bmatrix},$$

which gives $p + 1$ matrices in the group $PGL_2(\mathbb{F}_q)$. Let $S$ be the set of these matrices and define the Cayley graphs

$$X^{p,q} = \begin{cases} X(PGL_2(\mathbb{F}_q), S) \text{ if } \left(\frac{p}{q}\right) = -1 \\ X(PSL_2(\mathbb{F}_q), S) \text{ if } \left(\frac{p}{q}\right) = 1, \end{cases}$$

where $\left(\frac{p}{q}\right)$ is the Legendre symbol which equals 1 if $p$ is a square modulo $q$ and $-1$ otherwise. Then, the graphs $X^{p,q}$ form an infinite family of $p + 1$ regular Ramanujan graphs when we vary $q$ as shown in Lubotzky et al. (1988). The proof of this is not easy. We will sketch the main ideas.

To show that the graphs $X^{p,q}$ are Ramanujan, it suffices to show that the Ihara zeta function of these graphs satisfies the analogue of the Riemann hypothesis. Indeed, if we let $f_r$ to be the trace of the matrix $A_r$ in (12.4.1), we have for every $m$ (see p. 122 of Davidoff et al. 2003):

$$2 \sum_{0 \le r \le m/2} f_{m-2r} = \frac{2p^m}{n} \sum_{j=1}^{n} U_m\left(\frac{\lambda_j}{2\sqrt{p}}\right), \tag{12.5.1}$$

where $n = |X^{p,q}|$ and $U_m(x)$ is the Chebyshev polynomial of the second kind:

$$U_m(\cos\theta) = \frac{\sin(m + 1)\theta}{\sin\theta}$$

with $x = \cos\theta$. Using the theory of quaternions, the left-hand side of (12.5.1) is shown to be the number of ways of writing $p^m$ as the quadratic form:

$$a^2 + 4q^2(b^2 + c^2 + d^2).$$

Denoting this number by $s(p^m)$, one can use the theory of modular forms of weight 2 to show that

$$s(p^m) = \delta(p^m) + a(p^m),$$

where $\delta(p^m)$ is the coefficient of an Eisenstein series of weight 2 and $a(p^m)$ is the coefficient of a cusp form of weight 2 on $\Gamma(16q^2)$. (For a quick introduction to modular forms, we refer the reader to Ram Murty et al. 2015.) Now, one can show

$$\delta(p^m) = \begin{cases} 0 & \text{if } m \text{ is odd} \\ \frac{4}{n} \sum_{d|p^m} d & \text{if } m \text{ is even.} \end{cases}$$

This term corresponds to the *trivial* eigenvalue $\lambda_1 = p + 1$ appearing on the right-hand side of (12.5.1), and so we see that bounding the non-trivial eigenvalues of the $(p + 1)$-regular graphs $X^{p,q}$ is equivalent to estimating the cusp form coefficient $a(p^m)$. Keeping in mind that $q$ is fixed, the dimension of the space of cusp forms of weight 2 on $\Gamma(16q^2)$ is independent of $p$, and we see on invoking the Ramanujan conjecture (proved in this case by Eichler and Shimura) that the Ihara zeta function satisfies the analogue of the Riemann hypothesis. This is a cursory sketch of the argument, and the reader is referred to Davidoff et al. (2003), Lubotzky et al. (1988), Sarnak (1990) for the details.

We present now the main arguments of the result obtained by Marcus et al. (2015) that for any $k \geq 3$, there exists an infinite family of $k$-regular bipartite Ramanujan graphs. As we have seen earlier, one way to construct $k$-regular graphs is by using Cayley graphs of suitable groups and generating sets. Another way of constructing $k$-regular graphs is using the notion of 2-lift or 2-cover of a graph.

Consider a simple graph $X = (V, E)$. A 2-lift or a 2-cover of $X$ is a graph $Y = (V', E')$ obtained as follows. For each vertex $x$ of $X$, we create two distinct *copies* $(x, 0)$ and $(x, 1)$ of it in $Y$. The set $\{(x, 0), (x, 1)\}$ is sometimes called the fibre of $x$. For any edge $xy$ of $X$, we add either the pair of edges $\{(x, 0), (y, 0)\}$ and $\{(x, 1), (y, 1)\}$ or $\{(x, 0), (y, 1)\}$ and $\{(x, 1), (y, 0)\}$ to the graph $Y$.

We have an example in Fig. 12.1. The cycle $C_6$ is a 2-cover of the complete graph $K_3$. In general, a graph $X = (V, E)$ has $2^{|E|}$ 2-covers since for each edge $xy \in E$, there are two choices for the edges between the fibres of $x$ and $y$. The name cover is explained by the following proposition.

**Proposition 12.5.1** *Let $X = (V, E)$ be a graph. If $Y = (V \times \{0, 1\}, E')$ is a 2-cover of $X$, then the map $f : V \times \{0, 1\} \to V$, $f(x, j) = x$ from the vertex set of $Y$ to the vertex set of $X$ has the following properties.*

*(1) For any edge $\{(x, j), (y, \ell)\}$ of $Y$, $xy$ is an edge of $Y$.*
*(2) For any $(x, j)$ of $Y$, the function $f$ restricted to $\{(x, j)\}$ and its neighbourhood in $Y$ is a bijection between this set and the union of $\{x\}$ and its neighbourhood in $X$.*
*(3) If $X$ is $k$-regular, then $Y$ is $k$-regular.*

**Proof** The first part follows from the definition of a 2-cover: if $xy$ were not an edge of $X$, then we would not have the edge $\{(x, j), (y, \ell)\}$ in $Y$. For the second part, mapping $(x, j)$ to $x$ and any neighbour $(y, \ell)$ of $(x, j)$ in $Y$, to the neighbour $y$ of $x$ in $X$, creates a bijection between the two sets. The third part is a consequence of the second part. ∎

The last part of the previous result gives a recipe for constructing infinite families of $k$-regular graphs. Start with your favourite $k$-regular graph $X_0$. Take $X_1$ to be a 2-cover of $X_0$. By the previous proposition, $X_1$ has twice as many vertices as $X_0$ and is $k$-regular. Repeat the procedure for $X_1$. Take $X_2$ to be a 2-cover of $X_1$ and so on.

In order for this procedure to be useful for constructing Ramanujan graphs, one would like to be able to understand how the eigenvalues of the adjacency matrix of a 2-cover $Y$ of a graph $X$, relate to the eigenvalues of the adjacency matrix of $X$. It turns out that there is such a result and in order to state it, we need the important notion of signed adjacency matrix of a graph.

For a given graph $X = (V, E)$ and a sign function $s : E \to \{-1, 1\}$, **the signed adjacency matrix** $A_s$ **of** $X$ corresponding to the sign function $s$ is the $V \times V$ matrix where

$$A_s(x, y) = \begin{cases} s(x, y), & \text{if } x \sim y, \\ 0, & \text{otherwise.} \end{cases}$$

Equivalently, $A_s$ is obtained from the adjacency matrix $A$ of $X$ by replacing each non-zero entry $A(x, y)$ of $A$ from 1 to $s(x, y)$. If the sign function $s$ takes only value 1, then $A_s = A$. If $s$ takes only value $-1$, then $A_s = -A$. However, in other situations, there is no such a simple relation between $A_s$ and $A$. There is however a bijection between the 2-covers of $X$ and its signed adjacency matrices, which we explain below.

If $Y = (V \times \{0, 1\}, E')$ is a 2-cover of $X$, define the sign function $s : E \to \{0, 1\}$ as follows:

$$s(x, y) = \begin{cases} +1, & \text{if } (x, 0) \sim (y, 0) \text{ and } (x, 1) \sim (y, 1), \\ -1, & \text{if } (x, 0) \sim (y, -1) \text{ and } (x, 1) \sim (y, 0), \\ 0, & \text{otherwise.} \end{cases}$$

Note that one can reverse this procedure and from any sign function $s : E \to \{+1, -1\}$, one can construct a 2-cover of $X$.

In our example in Fig. 12.1, where $V = \{a, b, c\}$, the adjacency matrix and the signed adjacency matrix of $K_3$ are listed below:

$$A = \begin{bmatrix} 0 & +1 & +1 \\ +1 & 0 & +1 \\ +1 & +1 & 0 \end{bmatrix}, A_s = \begin{bmatrix} 0 & +1 & +1 \\ +1 & 0 & -1 \\ +1 & -1 & 0 \end{bmatrix}.$$

If the vertex set of the graph $C_6$ from Fig. 12.1 is listed in the order

$$\{(a, 0), (b, 0), (c, 0), (a, 1), (b, 1), (c, 1)\},$$

its adjacency matrix is below:

**Fig. 12.1** The cycle $C_6$ is a
2-cover of $K_3$

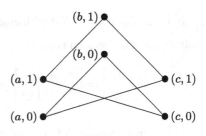

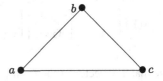

$$A(C_6) = \begin{bmatrix} 0 & 1 & 0 & 0 & 0 & 1 \\ 1 & 0 & 1 & 0 & 0 & 0 \\ 0 & 1 & 0 & 1 & 0 & 0 \\ 0 & 0 & 1 & 0 & 1 & 0 \\ 0 & 0 & 0 & 1 & 0 & 1 \\ 1 & 0 & 0 & 0 & 1 & 0 \end{bmatrix}.$$

We use this correspondence between 2-covers and signed adjacency matrices in the next result.

**Proposition 12.5.2** *Let $X = (V, E)$ be a graph with adjacency matrix $A$. If $Y$ is a 2-cover of $X$ whose associated sign function is $s$, then the spectrum of the adjacency matrix of $Y$ is the union of the spectrum of $A$ and the spectrum of the signed adjacency matrix $A_s$.*

*Proof* Consider the adjacency matrix $A(Y)$ of the graph $Y$, where the rows and columns are indexed in the order $(V \times \{0\}) \cup (V \times \{1\})$. This is the same as in the equation above, where $Y = C_6$. Writing

$$A(Y) = \begin{bmatrix} A_1 & A_2 \\ A_2 & A_1 \end{bmatrix},$$

where $A_1$ and $A_2$ are $V \times V$ symmetric matrices, observe that

$$A = A_1 + A_2,$$
$$A_s = A_1 - A_2.$$

If $u$ is an eigenvector of $A$ with eigenvalue $\theta$, then $\begin{bmatrix} u \\ u \end{bmatrix}$ is an eigenvalue of $A(Y)$ with eigenvalue $\theta$ as seen below:

$$A(Y)\begin{bmatrix} u \\ u \end{bmatrix} = \begin{bmatrix} A_1 & A_2 \\ A_2 & A_1 \end{bmatrix}\begin{bmatrix} u \\ u \end{bmatrix} = \begin{bmatrix} (A_1 + A_2)u \\ (A_2 + A_1)u \end{bmatrix} = \begin{bmatrix} Au \\ Au \end{bmatrix} = \theta \begin{bmatrix} u \\ u \end{bmatrix}.$$

If $v$ is an eigenvector of $A_s$ with eigenvalue $\tau$, then $\begin{bmatrix} v \\ -v \end{bmatrix}$ is an eigenvalue of $A(Y)$ with eigenvalue $\tau$:

$$A(Y)\begin{bmatrix} v \\ -v \end{bmatrix} = \begin{bmatrix} A_1 & A_2 \\ A_2 & A_1 \end{bmatrix}\begin{bmatrix} v \\ -v \end{bmatrix} = \begin{bmatrix} (A_1 - A_2)v \\ (A_2 - A_1)v \end{bmatrix} = \begin{bmatrix} A_s v \\ -A_s v \end{bmatrix} = \tau \begin{bmatrix} v \\ -v \end{bmatrix}.$$

Note that any vector of the form $\begin{bmatrix} u \\ u \end{bmatrix}$ is orthogonal to any vector of the form $\begin{bmatrix} v \\ -v \end{bmatrix}$.
If we take an orthogonal basis of $\mathbb{R}^V$ formed by eigenvectors of $A$ and an orthogonal basis of $\mathbb{R}^V$ formed by eigenvectors of $A_s$ and apply the arguments above, we can construct an orthogonal basis of $\mathbb{R}^{V'}$ formed by eigenvectors of $A(Y)$. Hence, the spectrum of $A(Y)$ is the union of the spectrum of $A$ and the spectrum of $A_s$.    ∎

In our example in Fig. 12.1, $X$ is the complete graph $K_3$ and the spectrum of $A$ is $2^{(1)}, -1^{(2)}$ (the exponents denote the multiplicities of the eigenvalues). We leave to the reader to figure that the spectrum of $A_s$ is $1^{(2)}, -2^{(1)}$ and that the spectrum of $C_6$ is $2^{(1)}, 1^{(2)}, -1^{(2)}, -2^{(1)}$.

Following the nomenclature used in Bilu and Linial (2006), the eigenvalues of $A(Y)$ that come from the spectrum of $A$ are called *old* eigenvalues and the eigenvalues of $A(Y)$ that are eigenvalues of $A_s$ are called *new* eigenvalues. A 2-cover $Y$ of $X$ *inherits* half of the spectrum of its adjacency matrix from the adjacency matrix $A$ of $X$ and the other half of *new* eigenvalues consists of the spectrum of the signed adjacency matrix $A_s$ of $X$.

The key breakthrough of Marcus et al. (2015) is the following theorem which partially solves a conjecture of Bilu and Linial (2006).

**Theorem 12.5.3** (Marcus et al. 2015) *Let $k \geq 3$ be a natural number and $X$ a connected $k$-regular graph.*

*(1) There exists a 2-cover $Y$ of $X$ such that any new eigenvalue of $Y$ is at most $2\sqrt{k-1}$.*
*(2) If $X$ is a bipartite graph, then there exists a 2-cover $Y$ of $X$ such any new eigenvalue of $Y$ has absolute value at most $2\sqrt{k-1}$.*

*Proof* We do not prove the first part here, but we note that the second part follows from the first since the spectrum of any bipartite graph is symmetric with respect to 0 (see Theorem 4.3.2) and the observation that any 2-cover of a bipartite graph must be a bipartite graph (which we leave as an exercise).    ∎

For any given $k \geq 3$, the following procedure yields an infinite family of $k$-regular bipartite Ramanujan graphs. Start with a bipartite $k$-regular Ramanujan graph $X_0$, say $X_0 = K_{k,k}$ the complete bipartite graph with $k$ vertices in each partite set. By part (2) of Theorem 12.5.3 and Proposition 12.5.2, there exists a 2-cover $X_1$ of $X_0$ such that $X_1$ is a bipartite $k$-regular Ramanujan graph. We then repeat this argument, and we deduce that there is a 2-cover $X_2$ of $X_1$ that is a bipartite $k$-regular Ramanujan. Continuing this process, one shows the existence of an infinite family $(X_m)_{m \geq 0}$ of bipartite $k$-regular Ramanujan graphs.

The signed adjacency matrix of a graph has many other interesting uses. We describe a recent application, Hao Huang's recent breakthrough proof of the Sensitivity conjecture from theoretical computer science (see Huang 2019). The Sensitivity Conjecture concerns the relationships between the sensitivity and block sensitivity of boolean functions (functions defined on the $n$-dimensional cube whose possible values are 0 or 1). It is equivalent to the following statement involving induced subgraphs of the $n$-dimensional cube. Recall that for a natural number $n$, the $n$-dimensional cube $Q_n$ has as vertices all the ordered $n$-tuples with entries 0 or 1 where two such $n$-tuples are adjacent if they differ in exactly one position.

**Theorem 12.5.4** *Let $Q_n$ denote the $n$-dimensional cube. If $X$ is an induced subgraph of $Q_n$ with $2^{n-1} + 1$ vertices, then $\Delta(X) \geq \sqrt{n}$.*

**Proof** There are two main ingredients for the proof of this theorem. The first ingredient relates the eigenvalues of a real and symmetric matrix to the eigenvalues of its principal submatrices. It is due to the French mathematician Augustin-Louis Cauchy (1789–1857) and is true in more generality for Hermitian matrices.

**Theorem 12.5.5** (Cauchy Interlacing) *Let $A$ be a real and symmetric $N \times N$ matrix with eigenvalues $\lambda_1 \geq \cdots \geq \lambda_N$. Let $B$ be a principal $M \times M$ submatrix of $A$ obtained by deleting the same set of rows and columns from $A$. Denote the eigenvalues of $B$ by $\mu_1 \geq \cdots \geq \mu_M$. The following inequalities are true for any $1 \leq j \leq M$:*

$$\lambda_j \geq \mu_j \geq \lambda_{j+n-M}. \tag{12.5.2}$$

The conclusion of the previous result is often referred to as stating that the eigenvalues of $B$ interlace the eigenvalues of $A$ and the previous result is commonly called Cauchy interlacing. We will not include a proof of this theorem here, and we refer the reader to Brouwer and Haemers (2012, Sect. 2.5).

The second ingredient is the following result, which is due to Huang (2019).

**Lemma 12.5.6** *The $n$-dimensional cube $Q_n$ has a signed adjacency matrix $B_n$ such that $B_n^2 = n I_{2^n}$. Consequently, the spectrum of $B_n$ is*

$$\sqrt{n}^{(2^{n-1})}, -\sqrt{n}^{(2^{n-1})}.$$

**Proof** The matrix $B_n$ will be constructed by induction on $n$. For $n = 1$, take $B_1 = \begin{bmatrix} 0 & 1 \\ 1 & 0 \end{bmatrix}$. Let $n \geq 2$ and assume that we have a signed adjacency matrix $B_{n-1}$ for $Q_{n-1}$

**Fig. 12.2** The
$n$-dimensional cube for
$n = 1, 2, 3$

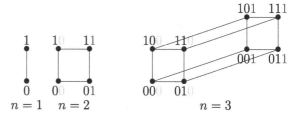

such that $B_{n-1}^2 = (n-1)I_{2^{n-1}}$. Note that $Q_n$ is constructed by taking two disjoint copies of $Q_{n-1}$, adding a 0 at the end of the vertices of one and 1 at the end of vertices of the other copy and finally, adding a perfect matching between these two copies. See Fig. 12.2 for an illustration.

With this construction in mind, define $B_n = \begin{bmatrix} B_{n-1} & I_{2^{n-1}} \\ I_{2^{n-1}} & -B_{n-1} \end{bmatrix}$. This is a legit signed

adjacency matrix of $Q_n$ and one can use the induction hypothesis $B_{n-1}^2 = (n-1)I_{2^{n-1}}$ to show that $B_n^2 = nI_{2^n}$. This gives the eigenvalues of $B_n$ and since the trace of $B_n$ is 0, we deduce that the multiplicities of $\sqrt{n}$ and $-\sqrt{n}$ are both $2^{n-1}$. ∎

Now we are all set to write the proof of Theorem 12.5.4. Let $X$ be an induced subgraph of $Q_n$ with $2^{n-1} + 1$ vertices. The principal submatrix $A_s$ of $B_n$ corresponding to the rows and columns indexed by the vertices of $X$ is a signed adjacency matrix of $X$. By Cauchy interlacing, its eigenvalues interlace the eigenvalues of $B_n$. Therefore, $\lambda_1(A_s) \geq \lambda_{1+2^{n-1}-1}(B_n) = \sqrt{n}$. As pointed to us by Willem Haemers, one can avoid using eigenvalue interlacing as follows. Because $A_s - \sqrt{n}I_{|X|}$ is a principal submatrix of $B_n - \sqrt{n}I_{2^n}$, we obtain that

$$\text{rank}(A_s - \sqrt{n}I_{|X|}) \leq \text{rank}(B_n - \sqrt{n}I_{2^n}) = 2^{n-1},$$

where the last equality follows from the spectrum of $B_n$. Since $|X| > 2^{n-1}$, this means that $A_s - \sqrt{n}I_{|X|}$ must be singular and thus, $\sqrt{n}$ must be an eigenvalue of $A_s$. This implies that $\lambda_1(A_s) \geq \sqrt{n}$ as desired.

By a similar argument to Corollary 6.4.7, one can show that $\Delta(X) \geq \lambda_1(A_s)$. Combining these two inequalities gives that $\Delta(X) \geq \sqrt{n}$. ∎

Huang's Theorem 12.5.4 states that no matter how one chooses more than half of the vertices in $Q^n$, there exists one vertex among them that has at least $\sqrt{n}$ neighbours from our set. In Fig. 12.2, no matter how one chooses 5 nodes in $Q^3$, one will be connected to at least $\lceil\sqrt{3}\rceil = 2$ others. While a case analysis may work for small $n$, such an approach would be hopeless for larger $n$ as the number of possible cases grows exponentially in $n$. Huang's theorem is the best possible in several ways. First, $2^{n-1} + 1$ cannot be replaced by anything smaller. The nodes of $Q^n$ can be split into two groups of size $2^{n-1}$ each: the nodes of even support on one side (000, 011, 101, 110 in $Q^3$ in Fig. 12.2) and the ones of odd support. There are no edges between nodes whose support has the same parity, so one could choose up to $2^{n-1}$ nodes of even support before any edge appears. Secondly, Chung et al. (1988)

showed in 1988 that there is an induced subgraph $X$ of $Q_n$ with $2^{n-1} + 1$ nodes such that $\Delta(X) = \sqrt{n}$ when $n$ is a perfect square.

The alphabet of computers has two letters/bits: 0 and 1 and their language consists of words over this alphabet. In theoretical computer science, researchers study boolean functions which can be thought of as boxes where one inputs binary words (typically of the same length) of 0s and 1s and the output is 0 or 1. Think about a form you fill where the questions have only yes/no answers. Your responses form a string of 0s and 1s (0 for no, and 1 for yes) and determine whether your form gets approved (output 1) or not (output 0). A more philosophical example is thinking about a given period in one's life and how each binary decision determined a particular outcome. Remember that $1 + 1 = 0$ for bits. Some examples of boolean functions are the two variable functions:

$$AND(x_1, x_2) = x_1 x_2$$
$$OR(x_1, x_2) = x_1 + x_2 + x_1 x_2.$$

Given a boolean function $f : \{0, 1\}^n \rightarrow \{0, 1\}$ whose inputs are all the binary words of length $n$ (for some natural number $n$) and a specific input $x = (x_1, \ldots, x_n)$ one can define the sensitivity $s(f, x)$ of the function $f$ with respect to the input $x$ as the number of entries $j$ where $f(x)$ changes when we flip the bit $x_j$ to $1 - x_j$ and leave the other bits unchanged. For example, $s(AND, 01) = 1$ as flipping the first bit in 01 changes the value of $AND(01)$ while flipping the second bit does not. In other words, $0 = AND(0, 1) = AND(0, 0) \neq AND(1, 1) = 1$. **The sensitivity of the function** $f$ is now defined as $s(f) := \max_x s(f, x)$. For our examples, $s(AND) = 2$ and $s(OR) = 2$.

The binary words of length $n$ can be interpreted as the nodes of the $n$-dimensional cube $Q^n$ seen earlier. Note that the sensitivity of a function $f$ with respect to a node $x \in Q^n$ is the number of neighbours $y$ of $x$ where $f(y) \neq f(x)$. The words $w$ such that $f(w) = 1$ induce a subgraph $X$ of the graph $Q^n$ while the words $z$ where $f(z) = 0$ induce the complementary subgraph $Q^n \setminus X$. For the AND function defined above, $X$ is the subgraph induced by 01, 10, and 11 while $Q^2 \setminus X$ is the just one vertex 00 (see Fig. 12.2 for $n = 2$).

In computer science, a lot of research is devoted to understanding how various sensitivity measures relate to each other. **The block sensitivity** of a boolean function $f : Q^n \rightarrow \{0, 1\}$ with respect to a given input $x$ is the maximum number of disjoint blocks $B_1, \ldots, B_k$ such that flipping the value of each of the entries whose index is in the set $B_j$ changes the value of $f(x)$ for any $1 \leq j \leq k$. The block sensitivity $b(f)$ is the maximum of $b(f, x)$ over all possible $x \in Q^n$. As an example, consider the function $g : Q^8 \rightarrow \{0, 1\}$ whose output is 1 on words of length 8 having 4 or 5 ones and 0, otherwise. Flipping any of the first 5 bits in $x = 00000111$ will change the value of $g(x) = 0$ and this leads to $s(g) = 5$. For $y = 11110000$, flipping the bits in either one of the disjoint blocks $\{1\}, \{2\}, \{3\}, \{4\}, \{5, 6\}, \{7, 8\}$ will change the value of $g(y) = 1$ giving $b(g) = 6$. It is not too hard to see that $s(f) \leq b(f)$ for any boolean function $f$ and a natural question is how well can one bound the block sensitivity in terms of the sensitivity. This was posed as a conjecture in 1992 by Nisan and Szegedy (1992) and became known as the sensitivity conjecture.

**Conjecture 12.5.7** (Sensitivity Conjecture) *There exists an absolute positive constant C such that for any boolean function f*

$$b(f) \le s(f)^C. \tag{12.5.3}$$

Several authors constructed boolean functions where the block sensitivity $b(f)$ is quadratic in the sensitivity $s(f)$, so if it exists, the constant $C$ in the above conjecture must be at least 2. It is known that any function $f : \{0, 1\}^n \to \{0, 1\}$ can be represented by a unique multilinear polynomial in $n$ variables, and we denote by $\deg(f)$ the degree of such polynomial. In 1992, Nisan and Szegedy proved that for any boolean function $f$,

$$b(f) \le 2 \deg(f)^2. \tag{12.5.4}$$

With (12.5.3) as a goal, a natural thing to do would be finding an upper bound for $\deg(f)$ in terms of the sensitivity $s(f)$. In 1992, Gotsman and Linial (1992) obtained an equivalence between such a result and a problem involving the cube $Q^n$. For any induced subgraph $X$ of $Q^n$, denote by $\Gamma(X) = \max(\Delta(X), \Delta(Q^n \setminus X))$, where $\Delta(X)$ is the maximum degree of $H$ and $\Delta(Q^n \setminus X)$ is the maximum degree of $Q^n \setminus X$.

**Theorem 12.5.8** (Gotsman and Linial 1992) *The following statements are equivalent:*

*(1)  For any boolean function $f : Q^n \to \{0, 1\}$, $\sqrt{\deg(f)} \le s(f)$.*
*(2)  For any induced subgraph $X$ of $Q^n$ with $2^{n-1} + 1$ vertices or more, $\sqrt{n} \le \Gamma(X)$.*

Gotsman and Linial conjectured that the above statements are true, and that is what Hao Huang proved in Theorem 12.5.4. Combining his result with (12.5.4), one gets that

$$b(f) \le 2s(f)^4, \tag{12.5.5}$$

proving the Sensitivity Conjecture with $C = 4$. There is still some work to do here, as the best examples known so far have a quadratic gap between $b(f)$ and $s(f)$.

The use of the signed $\pm 1$ adjacency matrices to represent graphs is not new and has been used for example since the 1960s work of van Lint and Seidel (1966) on equiangular lines. Huang's proof solved a 30-year-old conjecture and, in less than 1 year, has stimulated a lot of research activity. Alon and Zheng (2020) extended Huang's result and proved that it holds for a larger class of graphs. They also extended the $\pm 1$-signing of $Q^n$ and introduced unitary complex signings of graphs, where one can use $\pm 1$ and $\pm i$ as entries in a Hermitian adjacency matrix. Godsil et al. (2021) studied the connections with graph coverings and pointed out that the signed adjacency matrix of $Q^n$ constructed by Huang, appeared in 1985 in a paper by Cohen and Tits (1985) in a different context of finite geometries.

## 12.6 Exercises

**Exercise 12.6.1** Let $X$ be a connected $k$-regular graph with $n$ vertices and adjacency matrix eigenvalues $k = \lambda_1 > \lambda_1 \geq \cdots \geq \lambda_n$. Determine the eigenvalues of the complement $X^c$ of $X$.

**Exercise 12.6.2** The complete multipartite graph $K_{n_1,\ldots,n_k}$ is the graph whose complement is the disjoint union of the complete graphs $K_{n_1}, \ldots, K_{n_k}$. If $n_1 = \cdots = n_k$, determine its eigenvalues.

**Exercise 12.6.3** Let $k = \theta_0 > \theta_1 > \cdots > \theta_{d-1}$ be the distinct eigenvalues of the adjacency matrix $A$ of a $k$-regular connected graph $X$ with $n$ vertices. Show that

$$ J_n = \frac{n}{\prod_{i=1}^{d-1}(k - \theta_i)} \cdot \prod_{i=1}^{d-1}(A - \theta_i I_n). $$

**Exercise 12.6.4** A graph $X$ is **strongly regular** with **parameters** $(n, k, a, c)$ if it has $n$ vertices, is $k$-regular, every pair of adjacent vertices has $a$ common neighbours and every pair of non-adjacent vertices has $c$ common neighbours.

(1) Prove that $1 + k + \frac{k(k-a-1)}{c} = n$.
(2) Show that the adjacency matrix $A$ of $X$ satisfies the equation

$$ A^2 - (a - c)A - (k - c)I_n = cJ_n. $$

**Exercise 12.6.5** Let $X$ be a strongly regular graph with parameters $(n, k, a, c)$. Prove that the following statements are equivalent.

(1) The graph $X$ is not connected.
(2) $c = 0$.
(3) $a = k - 1$.
(4) The graph $X$ is a disjoint union of $t \geq 2$ copies of the complete graph $K_{k+1}$.

**Exercise 12.6.6** Let $X$ be a strongly regular graph with parameters $(n, k, a, c)$. It is called **primitive** if $X$ and its complement $X^c$ are both connected, and is called **imprimitive** otherwise. Prove that $X$ is imprimitive if and only if $X$ is either a disjoint union of two or more cliques $K_{k+1}$ or a non-empty complete multipartite graph $K_{b,\ldots,b}$.

**Exercise 12.6.7** Let $X$ be a primitive strongly regular graph with parameters $(n, k, a, c)$. Show that $X$ has three distinct eigenvalues. Determine them and their multiplicities in terms of $k, a, c$.

**Exercise 12.6.8** Let $q \equiv 1 \pmod 4$ be a power of a prime. The **Paley graph** $\mathbb{P}_q$ has vertices the elements of the field $\mathbb{F}_q$ with $x$ adjacent to $y$ if $x - y$ is a square in $\mathbb{F}_q$. Show that $\mathbb{P}_q$ is a strongly regular graph with parameters $\left(q, \frac{q-1}{2}, \frac{q-5}{4}, \frac{q-1}{4}\right)$. These graphs are named after the English mathematician Raymond Paley (1907–1933).

**Exercise 12.6.9** Let $n \geq 3$ be a natural number. Show that the line graph $\mathcal{L}(K_n)$ of the complete graph $K_n$, is a strongly regular graph and calculate its eigenvalues and multiplicities.

**Exercise 12.6.10** Let $n \geq 3$ be a natural number. The graph $H(n, 2)$ is defined as the line graph of the complete bipartite graph $K_{n,n}$. Prove that $H(n, 2)$ is strongly regular graph and find its spectrum.

**Exercise 12.6.11** Show that the Petersen graph is isomorphic to the complement of the line graph of $K_5$ and determine its spectrum.

**Exercise 12.6.12** Let $X$ be a connected and regular graph. If the adjacency matrix of $X$ has exactly three distinct eigenvalues, prove that $X$ is a strongly regular graph.

**Exercise 12.6.13** Let $k \geq 2$ be a natural number. Assume that $X$ is a strongly regular graph with parameters $(n, k, 0, 1)$.

(1) Prove that $n = k^2 + 1$.
(2) Calculate the eigenvalues and the multiplicities of $X$.
(3) Show that $k \in \{2, 3, 7, 57\}$.
(4) Prove that the cycle $C_5$ is a strongly regular graph with parameters $(5, 2, 0, 1)$. Prove that the Petersen graph is a strongly regular graph with parameters $(10, 3, 0, 1)$.

**Exercise 12.6.14** Let $X$ be the graph whose vertex set consists of the union of two collections of 25 pairs $(i, j) \in \mathbb{Z}_5 \times \mathbb{Z}_5$ and $(i, j)' \in \mathbb{Z}_5 \times \mathbb{Z}_5$. For each $i, j, k \in \mathbb{Z}_5$, $(i, j)$ is adjacent to $(i, j + 1)$, $(i, j + 2)'$ and $(j, ij + k)'$. Prove that the graph $X$ is strongly regular with parameters $(50, 7, 0, 1)$. This is **the Hoffman-Singleton graph**. It was constructed by the American mathematicians Alan Hoffman (1924–2021) and Robert Singleton in Hoffman and Singleton (1960).

**Exercise 12.6.15** A $n \times n$ matrix $C$ is called a **circulant matrix** if the row $i$ of $C$
· is obtained from the first row of $C$ by a cyclic shift of $i - 1$ steps for each $i \in [n]$. Let $Z$ be the $n \times n$ circulant matrix whose first row is $[0, 1, 0, \ldots, 0]$. Show that the eigenvalues of $Z$ are $1, \omega, \omega^2, \ldots, \omega^{n-1}$, where $\omega = \cos\left(\frac{2\pi}{n}\right) + i \sin\left(\frac{2\pi}{n}\right)$.

**Exercise 12.6.16** Let $C$ be a $n \times n$ circulant matrix whose first row is $[c_1, c_2, \ldots, c_n]$. Show that $C = \sum_{i=1}^{n} c_i Z^{i-1}$, where $Z$ is the $n \times n$ circulant matrix whose first row is $[0, 1, 0, \ldots, 0]$.

**Exercise 12.6.17** A **circulant graph** is a graph $X$ whose adjacency matrix is a circulant matrix. Show that a circulant graph is regular. If $[0, c_2, \ldots, c_n]$ is the first row of the adjacency matrix of $X$, show that its eigenvalues are $\lambda_s = \sum_{i=2}^{n} a_i \omega^{(i-1)s}$, for $s \in \{0, 1, \ldots, n - 1\}$ and $\omega = \cos\left(\frac{2\pi}{n}\right) + i \sin\left(\frac{2\pi}{n}\right)$.

**Exercise 12.6.18** Show that the cycle $C_n$ with $n$ vertices is a circulant graph and calculate its eigenvalues.

**Exercise 12.6.19  The Möbius ladder** $M_{2n}$ is the 3-regular graph on $2n$ vertices which is obtained from the cycle $C_{2n}$ by joining each pair of opposite vertices. Show that the Möbius ladder is a circulant graph. Show that the eigenvalues of the Möbius ladder $M_{2n}$ are

$$\lambda_s = 2\cos\left(\frac{\pi s}{n}\right) + (-1)^s,$$

for $s \in \{0, 1, \ldots, 2n - 1\}$.

**Exercise 12.6.20**  Recall from Exercise 4.6.18 that the **line graph** $\mathcal{L}(X)$ of a graph $X$ the edges of $X$ as vertices, two edges $e$ and $f$ of $X$ being adjacent in $\mathcal{L}(X)$ if they have common endpoint in $X$. Let $n \geq 3$ be a natural number. Determine which of the graphs $\mathcal{L}(K_n)$, $(\mathcal{L}(K_n))^c$, $\mathcal{L}(K_{n,n})$, $(\mathcal{L}(K_{n,n}))^c$ and $M_{2n}$ are Ramanujan.

# References

N. Alon, Eigenvalues and expanders. Combinatorica **6**, 83–96 (1986)

N. Alon, K. Zheng, Unitary signings and induced subgraphs of Cayley graphs of $\mathbb{Z}_2^n$. Adv. Comb. **2020**, Paper No. 11 (2020)

Y. Bilu, N. Linial, Lifts, discrepancy and nearly optimal spectral gap. Combinatorica **26**, 495–519 (2006)

A.E. Brouwer, W.H. Haemers, *Spectra of Graphs*. Universitext (Springer, New York, 2012)

F. Chung, Z. Füredi, R. Graham, P. Seymour, On induced subgraphs of the cube. J. Comb. Theory Ser. B **49**, 180–187 (1988)

F.R.K. Chung, Diameters and eigenvalues. J. Am. Math. Soc. **2**(2), 187–196 (1989)

S.M. Cioabă, On the extreme eigenvalues of regular graphs. J. Comb. Theory Ser. B **96**, 367–373 (2006)

A. Cohen, J. Tits, On generalized hexagons and a near octagon whose lines have three points. Eur. J. Comb. **6**, 13–27 (1985)

G. Davidoff, P. Sarnak, A. Valette, *Elementary Number Theory, Group Theory, and Ramanujan Graphs*. London Mathematical Society Student Texts, vol. 55 (Cambridge University Press, Cambridge, 2003), x+144 pp

P. Diaconis, M. Shahshahani, Generating a random permutation with random transpositions. Z. Wahrsch. Verw. Gebiete **57**, 159–179 (1981)

V.G. Drinfeld, The proof of Peterson's conjecture for $GL(2)$ over a global field of characteristic $p$. Funct. Anal. Appl. **22**, 28–43 (1988)

C. Godsil, M. Levit, O. Silina, Graph covers with two new eigenvalues. Eur. J. Comb. **93**, 103280 (2021), 12 pp

C. Gotsman, N. Linial, The equivalence of two problems on the cube. J. Comb. Theory Ser. A **61**, 142–146 (1992)

A.J. Hoffman, R.R. Singleton, On Moore graphs with diameters 2 and 3. IBM J. Res. Dev. **4**, 497–504 (1960)

S. Hoory, N. Linial, A. Wigderson, Expander graphs and their applications. Bull. Am. Math. Soc. (N.S.) **43**(4), 439–561 (2006)

H. Huang, Induced subgraphs of hypercubes and a proof of the sensitivity conjecture. Ann. Math. (2) **190**, 949–955 (2019)

A. Lubotzky, R. Phillips, P. Sarnak, Ramanujan graphs. Combinatorica **8**, 261–277 (1988)

A. Marcus, D. Spielman, N. Srivastava, Interlacing families I: bipartite Ramanujan graphs of all degrees. Ann. Math. (2) **182**, 307–325 (2015)

G. Margulis, Explicit group-theoretic constructions of combinatorial schemes and their applications in the construction of expanders and concentrators. (Russian), Problemy Peredachi Informatsii **24**(1), 51–60 (1988); translation in Probl. Inf. Transm. **24**(1), 39–46 (1988)

B.D. McKay, The expected eigenvalue distribution of a large regular graph. Linear Algebra Appl. **40**, 203–216 (1981)

B. Mohar, Isoperimetric numbers of graphs. J. Comb. Theory Ser. B **47**, 274–291 (1989)

M. Morgenstern, Existence and explicit constructions of $q + 1$-regular Ramanujan graphs for every prime power $q$. J. Comb. Theory Ser. B **62**, 44–62 (1994)

A. Nilli, On the second eigenvalue of a graph. Discret. Math. **91**, 207–210 (1991)

N. Nisan, M. Szegedy, On the degree of Boolean functions as real polynomials. Comput. Complex. **4**, 462–467 (1992)

M. Ram Murty, Ramanujan graphs: an introduction. Indian J. Discret. Math. **6**, 91–127 (2020)

M. Ram Murty, M. Dewar, H. Graves, *Problems in the Theory of Modular Forms*. IMSC Lecture Notes, vol. 1 (Hindustan Book Agency, 2015)

P. Sarnak, *Some Applications of Modular Forms*. Cambridge Tracts in Mathematics, vol. 99 (Cambridge University Press, Cambridge, 1990)

J.-P. Serre, Répartition asymptotique des valeurs propres de l'opérateur de Hecke $T_p$. J. Am. Math. Soc. **10**, 75–102 (1997)

J.H. van Lint, J.J. Seidel, Equilateral point sets in elliptic geometry. Indag. Math. **28**, 335–348 (1966)

# Chapter 13
# Hints

## 1. Basic Graph Theory

Exercise 1.6.1 Use Corollary 1.2.2.

Exercise 1.6.2 Use Theorem 1.5.2.

Exercise 1.6.3 Use induction on $n$.

Exercise 1.6.4 Use induction on $n$.

Exercise 1.6.5 Use the idea from Theorem 1.5.2.

Exercise 1.6.6 Consider a directed path of maximum length.

Exercise 1.6.7 Modify the proof of Theorem 1.4.1.

Exercise 1.6.8 Use Theorem 1.4.1.

Exercise 1.6.9 Determine the maximum number of edges in a complete bipartite graph with $n$ vertices.

Exercise 1.6.10 Each cycle $C_4$ must have two vertices of each colour.

Exercise 1.6.11 Partition the vertices of $Q_n$ according to the parity of their number of ones.

Exercise 1.6.12 Calculate the degrees of each vertex.

Exercise 1.6.13 Count the common neighbours of two given vertices.

Exercise 1.6.14 Use Exercise 1.6.13.

Exercise 1.6.15 Start with an arbitrary bipartite subgraph with two non-empty partite sets. For each vertex $x$, if the number of neighbours of $x$ which are contained in its colour class is greater than the number of neighbours of $x$ which are contained in the other colour class, then move $x$ to the other colour class.

Exercise 1.6.16 Use Theorem 1.4.1.

Exercise 1.6.17 Use proof by contradiction.

Exercise 1.6.18 Use proof by contradiction.

Exercise 1.6.19 Show first that $\sum_{xy \in E(X)} (d(x) + d(y)) = \sum_{z \in V(X)} d^2(z)$.

Exercise 1.6.20 Use Exercise 1.6.19 and the Cauchy-Schwarz inequality.

© Hindustan Book Agency 2022
S. M. Cioabă and M. R. Murty, *A First Course in Graph Theory and Combinatorics*,
Texts and Readings in Mathematics 55, https://doi.org/10.1007/978-981-19-0957-3_13

## 2. Basic Counting

Exercise 2.6.1 Calculate $\frac{\binom{n}{k+1}}{\binom{n}{k}}$.

Exercise 2.6.2 Use induction on $k$.

Exercise 2.6.3 Use induction on $k$ and the binomial formula.

Exercise 2.6.4 Use the formula $\binom{n}{k} = \frac{n!}{k!(n-k)!}$ or count the number of pairs $\{(K, L) : K \subset [n], |K| = k, L \subset K, |L| = l\}$ in two ways.

Exercise 2.6.5 Use the formula $\binom{n}{k} = \frac{n!}{k!(n-k)!}$ or count the number of $k$-subsets of $[n]$ depending on whether they contain $n$.

Exercise 2.6.6 Use Exercise 2.6.5.

Exercise 2.6.7 Count the number of $k$-subsets of $[m + n]$ in two ways.

Exercise 2.6.8 Use the formula $\binom{2n}{n} = \frac{(2n)!}{n!n!}$ and Stirling's formula for $n!$ and $(2n)!$.

Exercise 2.6.9 With binomial coefficients, if $|A| = k$, then there are $2^{n-k}$ subsets $B$ with $A \cap B = \emptyset$. For the combinatorial proof, consider the matrix whose rows are the characteristic vectors of $A$ and $B$.

Exercise 2.6.10 Use binomial formula for the first proof. For a bijective proof, if $n$ is odd, consider the function $A \rightarrow A^c$. If $n$ is even, use the fact that $n - 1$ is odd.

Exercise 2.6.11 If $\omega = \cos \frac{2\pi}{3} + i \sin \frac{2\pi}{3}$, calculate $(1 + \omega)^n$ in two different ways.

Exercise 2.6.12 Use $F_n = F_{n-1} + F_{n-2}$ and induction on $n$.

Exercise 2.6.13 Use a bijection between solutions and bracketings where $+1$ corresponds to an open bracket and $-1$ corresponds to a closed bracket.

Exercise 2.6.14 For the first part, write $n = 1 + 1 + \cdots + 1$ and construct a bijection between the solutions of the given equation and the $(k - 1)$-subsets of a set with $n + k - 1$ elements. For the second part, subtract one from each $x_i$ and use the first part.

Exercise 2.6.15 Use induction on $n$.

Exercise 2.6.16 Use binomial formula or count in two ways the number of triples $(A, x, y)$ where $A \subset [n]$ and $x, y \in A$.

Exercise 2.6.17 Use Stirling's formula.

Exercise 2.6.18 Let $x_k = \max\{x : \binom{x}{k} \leq n\}$ and use strong induction on $n$.

Exercise 2.6.19 Use the recurrence relation for the Bell numbers.

Exercise 2.6.20 Find a recurrence formula using the fact the last term of the sum can be 1 or 2.

## 3. The Principle of Inclusion and Exclusion

Exercise 3.6.1 Use Theorem 3.1.1.

Exercise 3.6.2 For the first part, a number is coprime with $p^{\alpha}$ if and only it is not divisible by $p$. For the second part, use the Chinese Remainder theorem.

Exercise 3.6.3 For $i \in [r]$, let $A_i = \{x : x \leq n, p_i | x\}$ and use Theorem 3.1.1. For another proof, use Exercise 1.6.2.

Exercise 3.6.4 Use Theorem 3.1.1.

Exercise 3.6.5 Use Theorem 3.1.1.

Exercise 3.6.6 Use induction on $n$.

Exercise 3.6.7 Use Sect. 3.2.

Exercise 3.6.8 Calculate $(f(t) - a_0)/t$ and $tf'(t)$.

Exercise 3.6.9 Use Theorem 3.4.2.

Exercise 3.6.10 The number of permutations with an even number of cycles is $|s(n, 2)| + |s(n, 4)| + \ldots$. The number of permutations with an odd number of cycles is $|s(n, 1)| + |s(n, 3)| + \ldots$.

Exercise 3.6.11 Use the counting idea from the proof of Theorem 3.5.3.

Exercise 3.6.12 Use inclusion and exclusion. If you have $m$ red cards and $n$ blue cards, count in two ways the number of $k$-elements subsets consisting only of red cards.

Exercise 3.6.13 The number $|s(n, 1)|$ equals the number of permutations with exactly one cycle. Use the recurrence relation.

Exercise 3.6.14 Use the definition of $S(n, k)$.

Exercise 3.6.15 Use induction on $n$ and the recurrence relations for $S(n, k)$ and $\binom{n}{k}$.

Exercise 3.6.16 Use Exercise 2.6.14.

Exercise 3.6.17 Calculate $(e^t - 1)g(t)$.

Exercise 3.6.18 Use Exercise 3.6.17.

Exercise 3.6.19 Use Exercises 3.6.17 and 3.6.18.

Exercise 3.6.20 Calculate $e^t f(t)$.

## 4. Graphs and Matrices

Exercise 4.6.1 Use Theorem 4.1.2.

Exercise 4.6.2 Use the definition of the Laplacian matrix.

Exercise 4.6.3 Multiply $A$ by the all one vector. Write down the eigenvector-eigenvalue equation for any eigenvalue and consider a vertex with the highest eigenvector entry. If $X$ is connected, prove that any eigenvector corresponding to $k$ has all entries equal.

Exercise 4.6.4 Write down the eigenvector-eigenvalue equation for any eigenvalue and consider a vertex with the highest eigenvector entry in absolute value. If $u$ is a vector, use the identity $u^T(kI_n + A)u = \sum_{xy \in E}(u_x + u_y)^2$.

Exercise 4.6.5 Use Theorem 4.1.2 to find the diagonal entries of $A$, $A^2$ and $A^3$.

Exercise 4.6.6 Evaluate tr $(A^2)$ in two ways and use the Cauchy-Schwarz inequality.

Exercise 4.6.7 If $x$ and $y$ have the same neighbours, consider the vector $u$ whose $x$th entry is $+1$, $y$th entry is $-1$ and is zero everywhere else.

Exercise 4.6.8 Use the definition of the eccentricity.

Exercise 4.6.9 Use the definition of the eccentricity.

Exercise 4.6.10 If $u$ is an eigenvector of $L(X)$ that is perpendicular to the all one vector, then prove that $u$ is an eigenvector of $L(X^c)$.

Exercise 4.6.11 For any vector $u$, use the identity

$$u^T Qu = \sum_{xy \in E}(u_x + u_y)^2.$$

Exercise 4.6.12 Prove that $\lambda_1$ is the maximum of $u^T A u / u^T u$, where the minimum is taken over all non-zero vectors $u$. For the other inequality, use a similar argument to Exercise 4.6.3.

Exercise 4.6.13 The eigenspaces of $K_n$ are the *one*-dimensional subspace spanned by the all one vector and its orthogonal complement.

Exercise 4.6.14 Consider at the coefficient of $\lambda^3$.

Exercise 4.6.15 Use Example 4.2.4.

Exercise 4.6.16 If $S$ is the set of distinct eigenvalues of $A$, consider the polynomial $h(x) = \prod_{\theta \in S}(x - \theta)$ and evaluate the matrix polynomial $h(A)$.

Exercise 4.6.17 The trace of $A^\ell$ equals the number of closed walks of length $\ell$ in the graph $X$.

Exercise 4.6.18 For any two edges $e$, $f$ of $X$, calculate the $(e, f)$-entry of the matrix $N^T N$.

Exercise 4.6.19 Use that $t^m \det(t I_n - N N^T) = t^n \det(t I_m - N^T N)$.

Exercise 4.6.20 Any eigenvalue of $N N^T$ is non-negative.

## 5. Trees

Exercise 5.6.1 Use Theorem 5.1.2.

Exercise 5.6.2 Use Theorem 5.1.2.

Exercise 5.6.3 Use induction on $n$ and the generalization of the binomial formula

$$(x_1 + \cdots + x_k)^m = \sum_{t_1 + \cdots + t_k = m} \frac{m!}{t_1! \cdot \ldots \cdot t_k!} x_1^{t_1} \cdot \ldots \cdot x_k^{t_k}.$$

Exercise 5.6.4 Use Theorem 5.1.2.

Exercise 5.6.5 For the lower bound, use Theorem 5.2.3. For the upper bound, draw the tree in the plane without crossing edges. Imagine that the edges are walls perpendicular to the plane and starting at the root, walk around this system of walls, keeping the wall always to your right. Each time we move away from the starting vertex, write a $+1$ and every time we move closer to the starting vertex, write a $-1$.

Exercise 5.6.6 Use Theorem 5.1.2.

Exercise 5.6.7 Use Theorem 5.2.3 and Exercise 5.6.1.

Exercise 5.6.8 Use the definition of a Prüfer code.

Exercise 5.6.9 The number of labelled trees on $n$ vertices is the same as the number of words of length $n - 2$ over an alphabet of size $n$.

Exercise 5.6.10 Use Sect. 5.2.

Exercise 5.6.11 Use Exercise 5.6.10.

Exercise 5.6.12 Use induction on $n$.

Exercise 5.6.13 Use Kruskal's algorithm.

Exercise 5.6.14 Use Exercise 5.6.4.

Exercise 5.6.15 Show that the maximum degree is 2.

Exercise 5.6.16 Consider a spanning tree of $X$.

Exercise 5.6.17 Use Theorem 5.3.2.

Exercise 5.6.18 Use Theorem 5.3.2.

Exercise 5.6.19 The number of vertices is less than the number of vertices of a $k$-regular tree of height $D$.

Exercise 5.6.20 Consider a path of maximum length in the tree.

## 6. Möbius Inversion and Graph Colouring

Exercise 6.6.1 Use the definition of a poset.

Exercise 6.6.2 Use the definition of the Hasse diagram and of the Möbius function.

Exercise 6.6.3 Use Theorem 6.1.7.

Exercise 6.6.4 Use Exercise 6.6.4.

Exercise 6.6.5 Use the definition of the Möbius function.

Exercise 6.6.6 Use the definition of a linear ordering.

Exercise 6.6.7 The constant term of $p_X(\lambda)$ equals $p_X(0)$. Use induction on the order of the graph for part (4).

Exercise 6.6.8 Use Theorem 6.3.2.

Exercise 6.6.9 Count the number of proper colourings of $Z$ by colouring $X$ first.

Exercise 6.6.10 Use Theorem 6.3.2.

Exercise 6.6.11 Every connected graph has a spanning tree.

Exercise 6.6.12 If the graph $X$ has $n$ vertices, prove that the coefficient of $\lambda^{n-1}$ in $p_X(\lambda)$ equals the negative of the number of edges in $X$.

Exercise 6.6.13 Use Theorem 6.3.2.

Exercise 6.6.14 Use the principle of inclusion and exclusion.

Exercise 6.6.15 Construct a graph whose vertices are the stations with two stations adjacent if and only if their distance is less than 150 miles.

Exercise 6.6.17 Use Theorem 6.4.9 to prove that $\chi(X) + \chi(\overline{X}) \leq n + 1$. Use the inequality $\chi(\overline{X}) \geq \omega(\overline{X}) = \alpha(X)$ to prove the second inequality.

Exercise 6.6.18 Partition the vertex set into $\chi(X)$ independent sets and count the edges between them.

Exercise 6.6.18 Combine a proper colouring of $X$ with a proper colouring of $Y$ to get a proper colouring of $X \vee Y$.

Exercise 6.6.20 For $1 \leq i \leq n - 2k + 1$, let $A_i$ denote the family of $k$-subsets whose smallest element is $i$. Show that $A_1, \ldots, A_{n-2k+1}$ and $V(K(n, k)) \setminus (A_1 \cup \cdots \cup A_{n-2k+1})$ are independent sets.

## 7. Enumeration under Group Action

Exercise 7.6.1 Use the definition of a group action.

Exercise 7.6.2 Use the definition of a group action.

Exercise 7.6.3 Use the definition of a group homomorphism.

Exercise 7.6.4 Use the definition of a group action.

Exercise 7.6.5 Use Pólya's Theorem.

Exercise 7.6.6 Use Pólya's Theorem.

Exercise 7.6.7 Use the definition of the cycle index polynomial.

Exercise 7.6.8 Use Pólya's Theorem.

Exercise 7.6.9 A graph with 4 vertices has at most 6 edges.

Exercise 7.6.10 Use the definition of the cycle index polynomial.

7.6.11 Use the results in the last section.

7.6.12 Use the results in the last section.

7.6.13 Use the definition of the cycle index polynomial.

7.6.14 Use the results in the last section.

7.6.15 Remember that an automorphism of $P_n$ is a bijection function $f : V(P_n) \to V(P_n)$ such that $xy \in E(P_n)$ if and only if $f(x)f(y) \in E(P_n)$.

7.6.16 Use the definition of the cycle index polynomial.

7.6.17 Use Pólya's Theorem.

7.6.18 Remember that an automorphism of a graph $X$ is a bijection function $f : V(X) \to V(X)$ such that $xy \in E(X)$ if and only if $f(x)f(y) \in E(X)$.

7.6.19 Use the definition of the cycle index polynomial.

7.6.20 Use Pólya's Theorem.

## 8. Matching Theory

Exercise 8.6.1 Use Hall's Theorem 8.1.1.

Exercise 8.6.2 Use the definition of orthogonal Latin squares.

Exercise 8.6.3 Use an argument similar to the proof of Theorem 8.3.2.

Exercise 8.6.4 Use induction on $t$.

Exercise 8.6.5 Use Proposition 8.2.1 and Exercise 8.6.4.

Exercise 8.6.6 Construct a bipartite graph with partite sets $\{A_1, \ldots, A_r\}$ and $\{B_1, \ldots, B_r\}$, where $A_i$ is adjacent to $B_j$ if $A_i \cap B_j \neq \emptyset$.

Exercise 8.6.7 Use the Hungarian algorithm.

Exercise 8.6.8 Use Tutte's Theorem 8.5.1.

Exercise 8.6.9 Use Tutte's Theorem 8.5.1.

Exercise 8.6.10 Use induction on $n$.

Exercise 8.6.11 Use Birkhoff-von Neumann Theorem 8.3.2.

Exercise 8.6.12 Add $t$ new vertices to $B$, join each of them to $A$ and use Hall's Marriage Theorem 8.1.1.

Exercise 8.6.13 Replace each vertex in $A$ by an independent set of $t$ vertices and use Hall's Marriage Theorem 8.1.1.

Exercise 8.6.14 Use Hall's Theorem 8.1.1 in the bipartite graph whose partite sets are the sets of rows and columns, respectively.

Exercise 8.6.15 If $X \setminus C$ is not a disjoint union of clique, then there are vertices $x, y, z, w$ such that $xy, xz$ are edges and $yz$ and $xw$ are not edges of $X$. Use the perfect matchings of $X \cup yz$ and $X \cup xw$ to construct a perfect matching for $X$.

Exercise 8.6.16 Use Tutte's Theorem 8.5.1.

Exercise 8.6.17 Consider the symmetric difference of two perfect matchings.

Exercise 8.6.18 Use induction on the number of vertices of $T$.

Exercise 8.6.19 Use Hall's Theorem 8.1.1.

Exercise 8.6.20 Follow the proof of Hall's Theorem 8.1.1.

## 9. Block Designs

Exercise 9.6.1 Use the results in Sect. 9.1.

Exercise 9.6.2 Use the results in Sect. 9.1.

Exercise 9.6.3 Use the results in Sect. 9.1.

Exercise 9.6.4 Use the definition of the Möbius function.

Exercise 9.6.5 Consider a design whose points are the students.

Exercise 9.6.6 Verify that any two points/edges are in exactly one block.

Exercise 9.6.7 Double count $(T, B)$, where $T$ is a subset of $t$ points, $B$ is a block and $T \subset B$.

Exercise 9.6.8 Prove that any two distinct vectors are in a unique triple.

Exercise 9.6.9 The row space of $AB$ is contained in the row space of $A$. The column space of $AB$ is contained in the column space of $B$.

Exercise 9.6.10 Use Theorem 9.3.1.

Exercise 9.6.11 Use Bruck-Ryser-Chowla Theorem 9.3.5.

Exercise 9.6.12 Use Bruck-Ryser-Chowla Theorem 9.3.5.

Exercise 9.6.13 Use the definition of a perfect code.

Exercise 9.6.14 Pick $d - 1$ coordinates and delete the corresponding entries in all the codewords.

Exercise 9.6.15 Use the definition of orthogonal Latin square.s

Exercise 9.6.16 Search for a subset $S$ such that $\mathbb{Z}_{n^2+n+1} \setminus \{0\} = \{s - s' : s \neq s' \in S\}$.

Exercise 9.6.17 Use Exercise 9.6.7.

Exercise 9.6.18 Prove that any two distinct elements of $\mathbb{Z}_n \times \mathbb{Z}_3$ are contained in exactly one triple.

Exercise 9.6.19 Any two distinct blocks have at most $t - 1$ points in common.

Exercise 9.6.20 Prove by contradiction.

## 10. Planar Graphs

Exercise 10.6.3 Find a plane embedding of the graph.

Exercise 10.6.4 Find a plane embedding of the graph.

Exercise 10.6.6 If $f_i$ denotes the number of faces of length $i$, then $2e = \sum_{i \geq g} i f_i \geq gf$.

Exercise 10.6.5 Find a cycle of the shortest length in the Petersen graph.

Exercise 10.6.1 Use the fact that $K_{2,n}$ is planar for any $n \geq 1$, but $K_{3,3}$ is not.

Exercise 10.6.2 Use Theorem 10.4.2.

Exercise 10.6.7 Show that $K_{4,4}$ without a perfect matching is isomorphic to the *three*-dimensional cube graph $Q_3$.

Exercise 10.6.8 Join any two points by an edge if and only if their distance in the plane is 1. Show this graph is planar.

Exercise 10.6.9 A non-planar graph has crossing at least 1.

Exercise 10.6.10 Use the definition of the crossing number and Theorem 10.1.2.

Exercise 10.6.11 Use Exercise 10.6.10 and find a drawing of $K_6$ with exactly 3 crossings.

Exercise 10.6.12 Use Exercises 10.6.6 and 10.6.10.

Exercise 10.6.13 Use the definition of outerplanar graphs. For $K_4$, assume it is outerplanar and derive a contradiction.

Exercise 10.6.14 Assume $K_{2,3}$ is outerplanar and derive a contradiction. Find a plane drawing of $K_{2,3}$.

Exercise 10.6.15 Use induction on the number of vertices.

Exercise 10.6.16 Use Exercise 10.6.15 and a greedy colouring.

Exercise 10.6.17 Decompose the polygon into triangles using non-crossing diagonals and use Exercise 10.6.16.

Exercise 10.6.18 A planar graph can be properly 4-coloured. Pick two colours and consider the subgraph induced by those colours. Do the same with the other two colours.

Exercise 10.6.19 Use induction on $n$.

Exercise 10.6.20 Write the set of solutions as the number of integer lattice points inside or on the boundary of a lattice triangle.

## 11. Edges and Cycles

Exercise 11.6.1 Use the definition of the line graph.

Exercise 11.6.2 Use proof by contradiction.

Exercise 11.6.3 Use proof by contradiction.

Exercise 11.6.4 Decompose the Hamiltonian cycle into two disjoint perfect matchings.

Exercise 11.6.5 Any cycle must alternate between the two partite sets.

Exercise 11.6.6 Start with a vertex and count the number of choices for the next edge of a Hamiltonian cycle.

Exercise 11.6.7 Use Fig. ??.

Exercise 11.6.8 Follow the proof of Theorem 11.2.3.

Exercise 11.6.9 Use the pigeonhole principle.

Exercise 11.6.10 Use the pigeonhole principle.

Exercise 11.6.11 Use induction and the pigeonhole principle.

Exercise 11.6.12 Use the pigeonhole principle.

Exercise 11.6.13 Modify the proof of Proposition 11.4.6.

Exercise 11.6.14 Use induction on $n$.

Exercise 11.6.15 Consider the polygon whose vertices are among the five points and which contains all the five points inside it.

Exercise 11.6.16 Use Exercise 11.6.15.

Exercise 11.6.17 Use Exercise 11.6.15 and induction on $n$.

Exercise 11.6.18 The neighbourhood of each vertex has no edges. Construct an independent set greedily.

Exercise 11.6.19 Use Proposition 11.4.6 and Exercise 11.6.18.

Exercise 11.6.20 Use induction.

## 12. Expanders and Ramanujan Graphs

Exercise 12.6.1 If $A$ is the adjacency matrix of $X$, then the adjacency matrix of its complement $X^c$ is $J_n - I_n - A$.

Exercise 12.6.2 Use Exercise 12.6.1.

Exercise 12.6.3 Consider an orthonormal basis for $\mathbb{R}^n$ formed by the eigenvectors of $A$.

Exercise 12.6.4 For a given vertex $x$, count the number of vertices at distance one and two from $x$, respectively. Calculate $A^2$ in terms of $I_n$, $A$ and $J_n$.

Exercise 12.6.5 Use the definition of a strongly regular graph.

Exercise 12.6.6 Use Exercise 12.6.5.

Exercise 12.6.7 Use Exercise 12.6.4.

Exercise 12.6.8 Let $\chi : \mathbb{F}_q \to \{-1, 0, +1\}$ be the quadratic character defined as

$$\chi(a) = \begin{cases} 0 & \text{if } a = 0, \\ +1 & \text{if } a = x^2 \text{ for some } x \neq 0, \\ -1 & \text{otherwise.} \end{cases}$$

Write and evaluate the number of neighbours of $a$ and $b$ as

$$\frac{\sum_{c \neq a,b} (\chi(c-a) + 1)(\chi(c-b) + 1)}{4}.$$

Exercise 12.6.9 Use the definition of a strongly regular graph.

Exercise 12.6.10 Use the definition of a strongly regular graph.

Exercise 12.6.11 Use Exercises 12.6.1 and 12.6.9.

Exercise 12.6.12 If $k > \theta > \tau$ are the distinct eigenvalues of the connected $k$-regular graph $X$ with adjacency matrix $A$, show that $(A - \theta I_n)(A - \tau I_n)$ is a multiple of the all one matrix $J_n$.

Exercise 12.6.13 Use Exercise 12.6.7.

Exercise 12.6.14 Use the definition of a strongly regular graph.

Exercise 12.6.15 Multiply the matrix $C$ by the column vector $[1, \omega^j, \omega^{2j}, \ldots, \omega^{(n-1)j}]^T$, for $j \in \{0, 1, \ldots, n-1\}$.

Exercise 12.6.16 The matrix $C$ can be written as the sum of $n$ circulant matrices. Use Exercise 12.6.15.

Exercise 12.6.17 Use Exercise 12.6.16.

Exercise 12.6.18 Use Exercise 12.6.17.

Exercise 12.6.19 Use Exercise 12.6.17.

Exercise 12.6.20 Use the definition of a Ramanujan graph and Exercise 4.6.19.

# Correction to: Graphs and Matrices

**Correction to:**
**Chapter 4 in: S. M. Cioabă and M. R. Murty,** *A First Course*
*in Graph Theory and Combinatorics,* **Texts and Readings**
**in Mathematics 55,**
**https://doi.org/10.1007/978-981-19-0957-3_4**

The original version of the book was inadvertently published with an incomplete
Figure 4.1 in Chapter 4, which has now been corrected. The book has been updated
with the change.

The updated version of this chapter can be found at
https://doi.org/10.1007/978-981-19-0957-3_4

© Hindustan Book Agency 2022
S. M. Cioabă and M. R. Murty, *A First Course in Graph Theory and Combinatorics*,
Texts and Readings in Mathematics 55, https://doi.org/10.1007/978-981-19-0957-3_14

# Index

© Hindustan Book Agency 2022
S. M. Cioabă and M. R. Murty, *A First Course in Graph Theory and Combinatorics*,
Texts and Readings in Mathematics 55, https://doi.org/10.1007/978-981-19-0957-3

Printed in the United States
by Baker & Taylor Publisher Services